THE DUSTY UNIVERSE

Photo: Babette S. Whipple

Fred Lawrence Whipple

THE DUSTY UNIVERSE

Proceedings of a symposium
honoring Fred Lawrence Whipple
on his retirement as Director of the
Smithsonian Astrophysical Observatory

October 17-19, 1973

George B. Field and
A. G. W. Cameron, Editors

Published for the
Smithsonian Astrophysical Observatory
by Neale Watson Academic Publications, Inc.
New York

Published for the
Smithsonian Astrophysical Observatory
by
Neale Watson Academic Publications
156 Fifth Avenue
New York, N.Y. 10010

Sole world distributor
excluding the United States, its possessions, and Canada
McGraw-Hill International Book Company

Library of Congress Cataloging in Publication Data
Main entry under title:

The Dusty universe.

Includes bibliographies and indexes.
CONTENTS: Field, G. B. and Cameron, A. G. W.
Introduction.--Cameron, A. G. W. The role of dust in
cosmogony.--Barshay, S. S. and Lewis, J. S. Chemistry
of solar material. [etc.]
1. Cosmic dust--Addresses, essays, lectures.
2. Interstellar matter--Addresses, essays, lectures.
3. Whipple, Fred Lawrence, 1906-   I. Whipple,
Fred Lawrence, 1906-   II. Field, George B., 1929-
III. Cameron, Alastair Graham Walter, 1925-
IV. Smithsonian Institution. Astrophysical Observatory.
QB791.D87   523.1'12   75-15576
ISBN 0-88202-033-1

Designed and manufactured in the U.S.A.

Table of Contents

INTRODUCTION

This book contains the proceedings of The Dusty Universe Symposium, held in honor of Professor Fred Lawrence Whipple on the occasion of his retirement as Director of the Smithsonian Astrophysical Observatory. The symposium was held in the Hilles Library of Radcliffe College, Cambridge, Massachusetts, October 17 to 19, 1973.

The general objective of the symposium was to survey the properties of interstellar and interplanetary dust and to explore possible cosmogonic relationships between the two. Research on these two subjects had tended to involve quite different observational techniques and quite different groups of people, and the organizers felt there was much to be gained by bringing these two groups together. Indeed, Fred Whipple himself suggested a cosmogonic relationship between the two as long ago as 1946.

Professor Whipple has had a long and distinguished career in astronomy. He was born in Red Oak, Iowa, in 1906. He obtained his A. B. and Ph. D. degrees at the University of California at Los Angeles and at Berkeley, respectively, and has subsequently been awarded a number of honorary degrees. After receiving his Ph. D. in 1931, he came to Harvard University, with which he has been associated ever since. Young men were expected to serve a longer academic apprenticeship in those days than is true now, so it was not until 1950 that he achieved the rank of Full Professor of Astronomy. He was appointed Phillips Professor of Astronomy in 1968, which chair he continues to hold today.

During World War II, as a research associate at the Radio Research Laboratory of the Office of Scientific Research and Development, he directed "Operation Window," which developed confusion reflectors or "windows" that were successfully used during the war to mislead enemy radar.

One of his major contributions as an astronomer was the development of a new technique for simultaneous photography of meteors from two or more stations. This

technique formed the basis for the first precise orbits of meteors and for a network of
16 automatic camera stations, which was established in the midwestern United States
for recording very bright meteors and for the recovery of meteoritic fragments. He
also carried out extensive research on comets and interplanetary dust, and he thus
became a leading authority on the properties of all types of small bodies within the
solar system.

In 1955, Dr. Whipple was appointed Director of the Smithsonian Astrophysical
Observatory. At this time, SAO moved from Washington, D. C., to Cambridge,
Massachusetts, where its proximity to the Harvard College Observatory made possible
a close and beneficial cooperation between the two observatories. The SAO had been
well known for its research in solar radiation and its effects on the earth. Dr. Whipple
enlarged this program to include studies of meteors, meteorites, the interplanetary
complex, and the upper atmosphere.

This formed the basis on which SAO was able to assume responsibility for the
optical tracking of American satellites launched during the International Geophysical
Year of 1957 to 1958. A Baker-Nunn tracking camera was designed and built to
Dr. Whipple's specifications, and 12 astrophysical observing stations equipped with
that camera were established around the world. The SAO also established a network
of amateur visual observers (Moonwatch), a computation center to receive the observational data and to prepare predictions of satellite transits, and a communications
network to link observing stations and teams with headquarters in Cambridge.

The scope of this program has been impressive. At the time of Dr. Whipple's
retirement as Director of SAO, the Baker-Nunn camera stations had made more than
400,000 successful photographs of satellite transits; from these films, the precise time
and position of 100,000 satellite images have been determined. The Moonwatch teams
recorded more than 250,000 observations. The resulting data have led to a much more
accurate representation of the gravitational field of the earth and an improved knowledge of the upper atmosphere. The precision of the Smithsonian tracking techniques
has recently been improved through the addition of laser tracking systems at several
of the camera sites.

During the period 1955 to 1969, Dr. Whipple planned and directed the growth of the Smithsonian Astrophysical Observatory from an initial staff of 5 to a peak staff of more than 500, of whom some 60 were scientists engaged in studies of the upper atmosphere, celestial mechanics, geodesy, gamma-ray astronomy, radio astronomy, meteoritics, stellar interiors, and comets and small particles in the solar system. Under Dr. Whipple's guidance and with NASA support, SAO developed one of the first orbiting observatories: Project Celescope, a telescope-television system to observe stars in the ultraviolet, which was launched aboard the OAO-2 satellite on December 7, 1968.

More recently, Dr. Whipple has directed the construction and development of a multipurpose astronomical observatory at Mt. Hopkins in Arizona. This is the largest and most extensive of the SAO field stations, and it will be the site of a new multiple-mirror telescope now under construction by SAO and the University of Arizona.

During his career, Professor Whipple has published more than 150 scientific papers on a wide variety of topics, ranging from stars and galaxies to the small bodies in the solar system; he has also written many popular articles. He has served in an advisory role on many different committees, both governmental and scientific, and has received a large number of honors and medals. A summary of his many activities is appended to this Introduction.

Dr. Whipple retired as Director of the Smithsonian Astrophysical Observatory on June 30, 1973. This has freed him to devote more time to his many research interests, which he continues to pursue vigorously. The subject of this symposium is one of the more active of those research interests, as is evident from the wide scope of his concluding remarks, which ended the symposium and which end this volume.

One of the highlights of the symposium was the dinner given in honor of Dr. Whipple, at which he was presented the Henry Medal, awarded by the Board of Regents of the Smithsonian Institution. This medal is given, on rare occasions, to individuals whose careers have combined unusual personal accomplishments in the arts or sciences with extraordinary service to the nation.

We wish to thank the other members of the local organizing committee, Mr. J. G. Gregory, Mr. R. G. Reed, Mr. J. C. Cornell, and Ms. K. Brownell, for their contribution to the success of the symposium. We also thank Ms. P. Brougham, Ms. P. Krauchunas, and Ms. J. Stepner for their assistance during the symposium. We have appreciated greatly the assistance of Ms. J. Copass with the editing of these proceedings.

George B. Field

A. G. W. Cameron

Center for Astrophysics

December 1974

Note: A new minor planet, numbered (1940), discovered at Harvard College Observatory's Agassiz Station on February 2, 1975, has been named for Dr. Whipple. The official citation reads:

"Named in honor of Fred L. Whipple, Harvard astronomer since 1931, Professor since 1950 and Director of the Smithsonian Astrophysical Observatory from 1955 to 1973. His countless contributions to our knowledge of the smaller bodies of the solar system include his icy-conglomerate model for cometary nuclei, and the development of modern techniques for the photographic observations of meteors. He has served as President of IAU Commissions 6, 15 and 22 and is now active on the NASA panel on space missions to comets and minor planets."

THE ROLE OF DUST IN COSMOGONY

A. G. W. Cameron
Center for Astrophysics
Harvard College Observatory and Smithsonian Astrophysical Observatory
Cambridge, Massachusetts

ABSTRACT

A discussion is given of the production of interstellar grain cores from stellar material, the gain and loss of grain mantles in interstellar space, chemical transformations in these grains when they become part of the primitive solar nebula, and the identification of these grains with interplanetary dust derived from comets and with the matrix material in meteorites. Thus, this paper proposes a common cosmogonic framework relating studies of interstellar and interplanetary grains.

1. INTRODUCTION

For a number of years, interstellar and interplanetary dust grains have been studied by quite different groups of people. However, it will be the principal theme of this paper that these two groups of studies form different facets of a single unified subject, at least in a cosmogonic sense.

Such a unified hypothesis does not constitute an entirely new idea. In 1946, Fred Whipple (see Whipple, 1948) suggested that an interstellar smoke cloud, as it was called in those days, participated in the formation of the solar system. He supposed that a cloud of interstellar gas was gradually condensing to form the sun, and that one of these interstellar smoke clouds came along and sideswiped the gas, giving it some rotation. He then supposed that the planets were formed from the smoke particles in the cloud. The smoke particles that came closer to the sun would lose their more volatile elements by evaporation; those remaining at greater distances from the sun would retain these volatile elements. In this way, there would be a characteristic chemical differentiation leading to terrestrial-type planets closer to the sun and giant planets with large amounts of volatile elements farther away from the sun. This was one of the earliest papers to recognize the importance of chemical reactions in the process of formation of the solar system. I do not believe these ideas of Whipple's have had the recognition they deserve.

More recently, Herbig (1970) has suggested that interstellar grains may be derived from material such as that which passed through the primitive solar nebula. He noted that the gaseous phase of the interstellar gas seems to be significantly depleted in the more refractory elements. Therefore, he suggested that much of the dust in the interstellar medium may have been formed in primitive stellar nebulae, similar to the primitive solar nebula that gave rise to our planetary system, and he assumed this dust could be expelled from the vicinity of the stars formed out of these primitive stellar nebulae by the operation of T Tauri stellar winds. As we shall see later, I think this process must also be of some importance.

2. GRAIN FORMATION

In normal interstellar space, the conditions are very unfavorable for the nucleation of interstellar grains, in view of the very low density of the gas and the presence of ultraviolet radiation, which makes it rather easy to break up molecular aggregates. A much more likely place to nucleate grains is in stellar atmospheres or in matter ejected from them.

In material of solar composition at pressures typical of stellar atmospheres, no significant amount of condensed matter can exist at temperatures above about 2000 K. Below that temperature, a characteristic sequence of condensations of different compounds occurs as the temperature is decreased. This has been explored in greatest detail by Grossman (1972), although for the somewhat higher pressures expected to be characteristic of the primitive solar nebula. However, since this sequence is far more sensitive to temperature than to pressure, the general condensation scheme should also hold for stellar atmospheres.

As matter of solar composition cools, the first compounds to be precipitated are those involving aluminum, calcium, and titanium oxides and silicates. These main constituents of condensed matter come out between 1800 and 1200 K. Below this temperature, nothing much happens until 680 K is reached, at which point sulfur will react with condensed metallic iron to form iron sulfide, or troilite. Since there is a greater natural abundance of iron than of sulfur, the excess metallic iron should be oxidized to form magnetite at a temperature of about 400 K. At still lower temperatures, the very abundant volatiles – water, ammonia, and methane – will become condensed.

This full condensation sequence is unlikely to be associated with grain formation in matter ejected from stellar atmospheres. Stellar-surface temperatures rarely fall below 2000 K, although grain formation in star spots or sunspots has been suggested by Hemenway (1974, this volume). Nevertheless, some small particles of very refractory materials may condense in stellar atmospheres and be ejected by radiation pressure.

There are a large number of astronomical scenarios in which matter is ejected from stellar surfaces. Sometimes, the ejection is quite gentle; sometimes, it is very violent. Perhaps the commonest kind of gentle ejection is in stellar winds. It might be thought that the gas densities in such winds would be far too small to allow grain formation, particularly since the radiation field of the star will keep the outgoing material at an elevated temperature as it expands. However, it is interesting to note that Geisel (1970) has found evidence for formation of dust particles under such conditions. She found there was an infrared excess accompanying the emission lines of Fe II and [Fe III]. The stars she has studied cover a wide range of spectral types, from B to M, but generally with some unusual emission characteristics that indicate more rapid rates of mass loss. Thus, it seems that dust formation is a common process accompanying stellar winds under a wide range of temperature conditions.

Most stars in the late stages of stellar evolution also appear to lose large amounts of mass through the formation of planetary nebulae. This mass is probably ejected in the form of a thick shell as a result of radiation pressure when the stellar core achieves a very high luminosity (Sparks and Kutter, 1974). It appears that this shell cools sufficiently during the ejection process so that water-vapor opacity becomes important. Thus, it is likely that grain formation occurs extensively throughout the ejected shell. However, when the ejected shell becomes somewhat more expanded, it becomes optically thin, and thus the radiation from the central star should heat the shell in its more expanded phases, presumably preventing condensation of the more volatile elements on the grain cores that have been formed. In terms of the amount of mass involved in stellar ejection processes, grain formation in planetary nebulae should thus be one of the most efficient forms for nucleation.

Another relatively gentle form of mass ejection occurs in stellar binary systems. In a closed binary system, when the more massive star reaches the giant phase of its evolution, it swells to fill its Roche lobe, and matter then spills over into the space surrounding the companion star. If the ejection velocity can become significant, then some of this mass is probably lost through one of the outer Lagrangian points and spirals away from the system. Some grain nuclei may be carried away by this gas flow; other grain nuclei that may be formed can possibly be ejected by radiation pressure. One binary system in which there may be evidence for this is ϵ Aur. I have suggested that the companion star in this system may be a black hole (Cameron, 1971), and probably the best model of the system is that constructed by Wilson (1971), who found that the companion star must be a compact object of high mass. In Wilson's model, a disk of dust surrounds the companion object, which is dense enough to transmit only about half the light, and there is an opening near the center of the disk where the dust density is very much less. It is evident that there are gas flows across the Roche lobe in this system, and it is very likely that the dust was formed in this way.

A more violent phenomenon in which dust formation appears to take place is the nova explosion. In such an event, about 10^{-4} or 10^{-5} of the stellar mass may be ejected. A particularly interesting set of observations of this phenomenon was made for Nova Serpentis 1970 (Geisel, Kleinmann, and Low, 1970). They observed an infrared excess that commenced a few days after the nova maximum; this excess became very strong a few weeks after the event.

On the other hand, it seems rather unlikely that we should expect dust formation to occur in supernova explosions. The velocity of the ejected envelope is so high, and the envelope temperature seems sufficiently high for weeks and months after the explosion, that grain nucleation probably remains unimportant until the ejected envelope gases can start sweeping up the surrounding interstellar material. The gases are then subjected to shock heating, which can very likely keep the temperature too high for grain formation until the material has become mixed with the interstellar medium. The only exception to this conclusion may involve the very small amount of mass in the outermost fringe of the star, which may be accelerated to extremely relativistic velocities. S. A. Colgate (private communication) has suggested that grain formation may occur because the temperature can become very low in the rest frame of this ejected material, and thus high-energy cosmic rays may in part consist of extremely relativistic grain nuclei. This negative conclusion is an important one if it is correct, because most of the radioactivities that should become important in considerations of cosmochronology should be made by nuclear reactions in such expanding envelopes, and therefore it seems likely that such radioactive material should be mixed into the interstellar medium without becoming incorporated in grain interiors.

Some stars manufacture enough carbon in the course of their evolution so that carbon exceeds oxygen in abundance in the surface layers, and the spectrum changes to that of a carbon star. Under these circumstances, the condensation sequence becomes quite different, with graphite and carbides forming the more refractory substances (O'Keefe, 1939; Gilman, 1969). When mass is lost from these stars, it is likely that grain nucleation will occur under circumstances similar to those involving matter of solar composition, but the grain nuclei will contain graphite and carbides.

It is important to recognize that the processes we have been discussing do not produce the interstellar grains in the form in which they are responsible for the extinction of interstellar starlight. The more volatile compounds have not been included in the nucleation process, only the more refractory ones. Much of the discussion of the properties of interstellar grains has involved core-mantle models of grains, where the core is some refractory material and the mantle is basically some form of dirty ice. Except in most unusual circumstances, we should therefore not expect to make a full core-mantle grain in the vicinity of a stellar surface by any of these processes, but only the cores of the grains, to which the mantles must be added later.

There are some interesting questions concerning the role of iron in these condensation processes. Since metallic iron and silicates are mutually immiscible (Blander and Katz, 1967), the meteoritic community has been debating whether these substances should nucleate and grow separately from one another in a condensation process. This question should also be considered in connection with the condensation processes we have been discussing, where the pressure is often orders of magnitude less than that which would be involved in the primitive solar nebula. If iron does nucleate and grow separately, then some interesting magnetic effects may occur in the vicinity of a stellar object. For very small sizes of the condensed nuclei, the metallic iron will exhibit a superparamagnetism, and when the grain growth reaches a certain critical size, around 200 μ, ferromagnetism becomes possible and the grain core may acquire a remanent magnetic field. At higher gas densities, the grain may condense and grow above the Curie point, and remanent magnetism can be acquired as a thermal remanent magnetism only when the grain cools below the Curie point. At rather low gas pressures, the iron core may nucleate and grow at temperatures below the Curie point, so that a chemical remanent magnetism becomes possible. If nucleation does not occur until extremely low gas pressures are reached, then the iron cores will nucleate and grow in the form of magnetite. In the presence of a magnetic field of order 1 G, the magnetite cores will acquire a significant chemical remanent magnetism. These effects will not be significant in terms of the alignment of interstellar grains in interstellar space by interstellar magnetic fields, but they will prove to be interesting possibilities in connection with our later discussion of dust in the solar system.

Meanwhile, it is interesting to note that Huffmann (1973) has found that small particles of magnetite reproduce very well many of the polarization effects that have been observed for interstellar dust extinction. If this can be taken as an indication that magnetite cores are common in interstellar grains, then it will become an interesting problem to decide whether the magnetite was formed by oxidation of iron cores or whether it was formed by direct condensation at very low gas pressures. The oxidation of metallic iron to magnetite can take place only at a temperature of about 400 K and lower. However, the temperature should presumably not be too much lower than 400 K, or the chemical reaction rates would become too slow to perform this transformation. If magnetite is very common in interstellar grains, a predominant formation by chemical conversion of metallic iron would imply that astrophysical scenarios where gas is held at temperatures slightly below 400 K for considerable lengths of time should be very common. Since this seems to me to be somewhat unlikely, I prefer to think that the magnetite was probably mostly formed by condensation processes at extremely low gas pressures.

3. GRAINS IN THE INTERSTELLAR MEDIUM

In the discussion so far, we have seen a number of ways in which grain nuclei can be injected into the intersteller medium. These grain nuclei will consist of very refractory materials, but the processes of condensation are unlikely to allow a significant condensation of the more volatile elements, and not all the refractory materials — particularly those ejected from supernovae — should be tied up in grain nuclei upon injection into the interstellar medium.

Once the grain nucleus is in the interstellar medium, it is subjected to an environment that is very far from local thermodynamic equilibrium. The gas kinetic temperature will be of order 10^2 K if in an H I region, or approaching 10^4 K if in an H II region. The internal excitation temperature of the grain nuclei will be much below these temperatures, and the dilute ultraviolet photons will have temperatures of order 10^4 K. The grains will also be bombarded by superthermal particles in the form of cosmic rays. Hence, their subsequent behavior represents a somewhat complicated physical problem, but in general there will be competition between grain growth and grain depletion.

The process of grain growth consists of the addition to the grain of both refractory and volatile atoms, but excludes hydrogen and the rare gases. Under typical interstellar conditions, such atoms will occasionally collide with a grain surface and will stick, at least temporarily, to the surface. In addition to the weak binding provided by van der Waals' forces, if any stronger chemical bonds are available from the neighboring atoms, these are likely to be established. Absorption of photons by the grain is likely to lead to internal degrees of excitation that will promote the establishment of such bonds. Because of the stochastic fashion in which the resulting material was assembled, one can expect a veritable chemical zoo of different chemical compounds, the resulting mixture being very exotic. Because of the large abundance of oxygen and hydrogen, water should be a very abundant constituent of this zoo, but nevertheless a majority of the oxygen is likely to be tied up in other types of chemical compounds. This is an extreme form of what Fred Whipple would call "dirty ice." This chemical zoo should constitute what we can call the grain's mantle, and if the growth of the grain continues until there is a depletion of volatile elements in the surrounding region of interstellar space, then the mass of the mantle will be considerably greater than the mass of the grain core.

Some aspects of this growth problem have been discussed by Watson and Salpeter (1972), who have also dealt with the various possible destruction mechanisms for the mantle. They have concluded that photosputtering should be the most important destruction mechanism, involving ultraviolet photons in the interstellar-starlight radiation field. Such destruction depends sensitively on the nature of the chemical bonds in the mantle; for example, sputtering is inefficient if the mantle should form a crystalline lattice. Aannestad (1973) has suggested that sputtering by impact of gas atoms, particularly helium, may also be important, under some circumstances, in the destruction of grain mantles. The particular case discussed by him involves an interstellar gas-shock wave, which must be considered relatively rare in most regions of interstellar space at any given time.

We may now state the generality that grain growth will be favored in those parts of the interstellar medium that are of higher density, so that the rate of collisions of interstellar gas atoms with grain surfaces will be correspondingly increased. The photosputtering destruction process will be largely independent of the gas density; in very dense regions, the rate will be decreased by shielding, and, in H II regions, the rate may be increased because a larger part of the ultraviolet spectrum is available for the process.

This leads to a picture for the behavior of the interstellar grains during their residence in the interstellar medium that I base largely on papers by Mészáros (1972, 1973a, b). We adopt a density-wave picture of the galactic spiral arms. As gas flows into such an arm, it undergoes some shock compression (Roberts, 1969; Roberts and Yuan, 1970), followed by the Parker (1966) instability, which allows gas to flow down interstellar magnetic-field lines to form a new set of interstellar clouds. Since the cloud density is locally higher than average, the rate of growth of the mantles on the grains imbedded in it is likely to exceed the rate of destruction. The growth of the grain mantles will thus tend to deplete the majority of the atomic species contained within the cloud. This process eliminates those atoms and ions that play a large role in the cooling of the gas, so that the gas cloud will gradually be heated as its cooling efficiency declines. As the cloud warms, it expands and becomes less easily detectable. Eventually, the cloud will revert to the higher temperature state characteristic of the intercloud medium. This may be hastened by the heating that occurs as a result of collisions between the clouds. When the the gas emerges from the galactic arm, its density is likely to be low and the temperature, quite high. Under these conditions, photosputtering probably once again predominates, and the grain mantles are gradually eroded. Whether there remains some remnant of the cloud structure in the interarm region is as yet an unsettled question; in any case, the cloud densities, if there are any, should be very low, and the majority of the volatile elements in the grain mantles should be restored to interstellar space. However, because of the photochemical selectivity of the photosputtering process, it is possible that certain of the elements are retained in a more tightly bound inner mantle adjacent to the core and do not revert to the interstellar medium. This may be the cause of the very strong general depletion of certain elements in the interstellar medium, such as calcium. Clearly, a lot of interesting experimental and theoretical work in photochemistry will be needed to settle such questions.

It is my own belief that the optimum condition for star formation occurs at the time of the interstellar cloud formation. In the more massive clouds, the internal pressure may not be sufficient to offset the effects of the self-gravitational field of the cloud, and the cloud may go right on into collapse, forming an association of stars as a result of subsequent fragmentation. Because of the high density and low temperature in such collapsing clouds, the volatile elements should very rapidly collect onto the grain nuclei and completely deplete the gas phase of these elements. Thus, in the early phases of interstellar collapse forming stars, the accompanying grains should have fully developed mantles.

It is interesting to speculate whether grains having these properties can reproduce the observed properties of interstellar extinction (Aannestad and Purcell, 1973). There is a characteristic feature in the extinction curve that has been interpreted as being due to graphite. Since some of the grain nuclei are expected to be composed of graphite, this feature receives a natural explanation. It is also well known that a significant fraction of the grains become aligned by the fairly weak interstellar magnetic field. This process should certainly be assisted by those grain cores composed of magnetite or metallic iron. As mentioned before, magnetite also possesses suitable polarization properties for the interstellar grains. Thus, there seems to be a general consistency of at least these gross properties of the interstellar extinction.

4. THE PRIMITIVE SOLAR NEBULA

We are now ready to turn to the role played by interstellar grains in the formation of the solar system. For this, we need to know the type of physical conditions to be expected throughout the primitive solar nebula formed at the time of formation of the solar system.

There are a great many ideas about the character of the primitive solar nebula, and space does not permit a discussion of the differences between these. I shall base the following discussion on my own ideas in this regard (Cameron and Pine, 1973; Cameron, 1973a), not only because I have a natural prejudice in favor of them, but also because the resulting model of the primitive solar nebula is the only one in which an attempt has been made to calculate the parameters of the nebula from the inter-stellar-collapse conditions. Without the constraints imposed by these conditions, it would be very difficult to make any quantitative statements about the relationship between the interstellar grains and dust in the solar system.

Since the derivation of the models has been extensively discussed in the literature, I shall mention only a few salient points here. I expect that the collapsing interstellar gas will be highly turbulent, and this leads to two conclusions. The first is that the total angular momentum of the primitive solar nebula should result from the random motions of the turbulent eddies in the collapsing gas, which leads to a calculated value for this total angular momentum. The other conclusion is that the turbulent motions of the gas should promote collisions among the interstellar grains contained in the gas; since the grain mantles are probably rather fluffy structures, this may lead to the grains sticking together. In turn, these grain aggregates should stick to other grain aggregates, and fairly sizable "snowballs," up to several centimeters in characteristic dimension, may be built up as a result of these mutual collisions by the time the gas approaches its final place in the primitive solar nebula.

Given the total angular momentum estimated in this way, and assuming a total of two solar masses of gas in the primitive solar nebula, there still remains a choice concerning the distribution of the total angular momentum as a function of the radial distribution of the collapsing mass. It turns out that this distribution does not matter a great deal, as long as it is reasonable, as can be seen in Figure 1. This figure shows two possible surface densities of the gas in the primitive solar nebula, corre-sponding to two choices of the angular-momentum distribution, labeled "uniform" and "linear," respectively. It should be noted from Figure 1 that the primitive solar nebula contains only about 10^6 g cm^{-2} at the center, so that all the mass is spread out and there is no central concentration of mass that can be identified with the sun. The sun must form from the primitive solar nebula as a result of gaseous dissipation processes occurring within it.

Another constraint related to the interstellar collapse concerns the thermodynamics
of the initial solar nebula. Because the cooling efficiency of the collapsing gas becomes
very great, and because the internal heating becomes quite small, the temperature
should fall to about 5 or 10 K during the early collapse process. The temperature will
be maintained at about this level until the gas becomes sufficiently dense to be able to
absorb its own emitted infrared radiation. At this stage, the collapse changes from
being isothermal to adiabatic; the gas is eventually heated adiabatically, so that higher
temperatures accompany higher pressures toward the center of the primitive solar
nebula. The ultimate compressive adiabat was estimated in numerical calculations by
Larson. One can also get an estimate of it by use of the cosmothermometers and
cosmobarometers derived from meteorite assembly conditions by Anders and others.
These two very different approaches give comparable values for the compressive
adiabat, and the combinations of temperature and pressure for the primitive solar-
nebula models, which are shown in Figure 2, are based on these considerations.

It is necessary to make some cautionary remarks about the distance scale shown
in Figures 1 and 2. Because the mass in the primitive solar nebula is greatly spread
out, the gravitational potential has a much smaller gradient in the nebula than in the
present solar system, where the majority of the mass is strongly concentrated toward
the center. Therefore, a massive body formed at some radial distance in the primitive
solar nebula will acquire a different radial distance after the bulk of the gas has flowed
past that body to form the sun and the gravitational-potential gradient has increased.
Thus, Mercury, now at a radial distance of just less than 0.4 a.u., should be formed
at 2 or 3 a.u.; the asteroid belt should be formed around 9 a.u. However, Uranus
and Neptune should not be formed very much farther out than their present distances,
since the bulk of the solar nebula is at a smaller radial distance than their locations are.
It can be seen that the thermodynamic conditions near 9 a.u. in the primitive solar
nebula are consistent with the meteoritic cosmothermometers and cosmobarometers.

A typical structure of the resulting primitive solar nebula is shown in Figure 3.
The two hatched regions represent places where energy transport occurs by convection,
carrying energy from the interior toward photospheric surfaces from which it can be
radiated into space. The inner convective regions result from the properties associated
with the dissociation of hydrogen molecules. Since the temperature in this region

exceeds 2000 K, no condensed solids will be present there. The outer hatched region represents convection that is driven as a result of the high opacity associated with metallic iron. Outside the hatched regions, energy transport is by emission and absorption of radiation, and the only large-scale gas motions will be those associated with the dissipation of the primitive solar nebula.

These large-scale gas motions arise from the fact that a steady state cannot be achieved in a differentially rotating flattened object such as this, since it is not possible to have complete coincidence of surfaces of constant pressure, density, and temperature. Therefore, large-scale circulation currents are set up, in which gas tends to flow outward at high altitudes away from the central plane of the nebula. This transports angular momentum outward, and there is an accompanying mass flow inward to form the sun.

A fundamentally important point is that the interstellar grains that participated in the formation of the primitive solar nebula cannot be completely evaporated until temperatures exceed 1800 K. Such temperatures are achieved only at distances of less than 2 a.u. Beyond this distance, all condensed matter must be formed from the remnants of the interstellar grains. This is a very general conclusion, which depends only in a minor way on the details of the model construction. Collapsing interstellar gas will cool to a very low temperature, and only at high gas densities can gas compression heat the gas sufficiently to destroy these grains. One should expect to achieve the necessary temperatures beyond distances of the order of 2 a.u. only in extremely massive bodies, which are not of interest in the present context. Thus, the survival of interstellar grains, in at least some parts of their cores, is to be expected in the outer parts of the primitive solar nebula. However, the grains will have lost their dirty-ice mantles at distances of less than something like 15 a.u. The more volatile components of the dirty-ice mantles will be lost at considerably greater distances.

If a significant number of the interstellar grains are clumped together into bodies of the order of centimeters in dimension at the time of formation of the primitive solar nebula, and if the lacy framework of stone and iron survives the loss of the volatile mantles, then these bodies will descend fairly rapidly toward the central plane of the nebula. Goldreich and Ward (1973) have recently shown that this fairly thin layer of condensed material should become gravitationally unstable toward a concentration into bodies of typically asteroidal mass. This is a first major step toward the formation of planetary bodies. This process will be greatly impeded by the outer convection zone in the primitive solar nebula, so that we should expect much more rapid formation of planetary bodies beyond the outer convection zone.

It must be expected that not all the collapsing interstellar gas will manage to become a part of the primitive solar nebula. Because of the turbulent motions, some of it will have enough transverse momentum to form separate condensed gaseous bodies in orbits about the primitive solar nebula itself. We call these satellite nebulae. They should have considerably smaller masses than does the primitive solar nebula itself, so that the gas pressures in them will be considerably less and the temperatures will remain quite low throughout.

This sets the stage for our subsequent discussion of the behavior of the interstellar grains within the context of a solar-nebula model such as that described.

5. COMETS

It has become generally accepted that an extended cloud of comets surrounds the solar system, with typical aphelion distances of the order of 10^5 a.u. The manner in which this Oort (1950) cloud was originally established will dictate the relationship of the interstellar grains to the composition of comets. It has also become generally accepted that comets are giant dirty snowballs as originally suggested by Fred Whipple (1950). The "dirt" was originally intended to mean solid dust grains imbedded within a matrix of water ice. However, the word "ice" should also be considered, in a very general sense, to include any of the very volatile gases that are frozen into the composition of the comet. Which volatiles are present will also depend on the mode of formation of the comets.

If the comets were originally made in the asteroid belt and ejected to large distances through the gravitational action of Jupiter, as postulated by Oort and recently advocated again by Öpik (1973), then some interesting boundary conditions can be applied from chemical considerations. If we believe that meteorites have come from the asteroid belt, then meteoritic cosmothermometers indicate that ordinary chondrites were assembled in a gas at a temperature of about 450 K and carbonaceous chondrites at about 350 K (Anders, 1972). This gives an indication of the temperature achieved by the primitive solar nebula in at least its initial stages in the region from which the asteroids developed. Under such conditions, the interstellar grains would have lost their volatile gases, but they would otherwise have been little affected. Recent work by Podolak and Cameron (1974) suggests that Jupiter contains a considerable amount of water in excess of solar composition. This work, together with that of Perri and Cameron (1974), suggests that Jupiter was formed only after a very large core of condensed materials had accumulated into a planetary body and the solar nebula had cooled substantially from its initial temperature conditions until a dynamic collapse onto the condensed planetary body became possible. Thus, it would seem that comets could be formed in the asteroid belt or in the vicinity of Jupiter only in the later stages of planetary accumulation. Not only would the exotic chemical zoo composing the mantles of the interstellar grains have evaporated under the initial conditions in the solar nebula in this vicinity, but the volatile gases would have reacted chemically to become much simpler compounds, mainly H_2O, NH_3, and CH_4, although nonthermal processes may have been responsible for the production of a wider variety of carbon compounds. However, it seems unlikely that compounds much more volatile than water would have condensed during the cometary formation process. Thus, the Jupiter ejection hypothesis requires a relatively simple icy composition for the comets, with imbedded dust grains.

If comets were formed in the vicinity of Uranus and Neptune and ejected into the Oort cloud by these planets, then again it is likely that a considerable chemical simplification of the volatile materials would have taken place. The difficulty with this hypothesis is that the mass of comets in the Oort cloud is probably considerably greater than the masses of Uranus and Neptune together, and the situation is even worse if one allows for the comets that would be ejected from the solar system altogether; thus, why should not Uranus and Neptune be driven farther into the inner solar system by such an ejection process?

I have recently suggested (Cameron, 1973a) that the comets were formed in the satellite nebulae, described in the preceding section, postulated to be in orbits that locate them in the vicinity of the Oort cloud. Temperatures should remain low throughout such satellite nebulae, and comets should be formed throughout them by the collecting together of interstellar grains. I would expect that only a minor degree of chemical simplification of the volatile compounds should have occurred in most of the comets formed under these circumstances.

All these possibilities have in common the feature that the dust grains imbedded in the dirty snowball should be the cores of the interstellar grains, and these cores should be essentially unaffected by the process of cometary formation. The pictures differ in the degrees to which the volatile mantles of the original interstellar grains can be expected to be chemically transformed by heating during cometary formation. The farther out the postulated site of cometary formation, the less the resulting chemical transformation is, and the greater is the preservation of the original exotic chemical zoo that the mantles of the interstellar grains should possess.

Observations of such molecules as C_3 and CH_3CN in comets suggest that a considerable degree of chemical complexity is present. However, analysis of the situation is far from simple. Bombardment by cosmic rays over several billion years can cause a considerable chemical complexity in the upper few meters of a new comet. The more volatile gases will be vaporized and will diffuse out of the icy cometary matrix as the temperature of the cometary surface is increased, and thus these substances will be predominantly lost at fairly large distances from the sun, where the comet is difficult to observe spectroscopically. When a comet comes close to the sun, the gas temperature in the coma surrounding it can become quite high, thus promoting extensive chemical reactions among the constituents present. We lack a quantitative analysis of these effects, but perhaps the large body of observational data being acquired in connection with Comet Kohoutek will stimulate some progress in this direction.

It is well known that the bulk of the meteors derived from interplanetary dust are cometary in origin. The present discussion strongly suggests that these meteors are composed of the cores of interstellar grains. However, most meteors are very much more massive than the cores of interstellar grains, and it is thus evident that they must represent clusters of such grain cores. Individual grain cores would presumably be ejected from the solar system by radiation pressure.

It is interesting that analysis of meteoric light curves in the atmosphere indicates that the meteors have a very low mean density. They must consist of a fragile lacy framework of interstellar grain cores stuck together. This picture has been confirmed from the collection of meteoric dust at high altitudes in rocket experiments by Hemenway and his collaborators (1974, this volume). The meteoric dust particles are very fragile and tend to be filamentary in structure.

It is not surprising that snowballs made of interstellar grains should produce such fragile rocky structures. If the volatile gases evaporate away in a fairly gentle fashion, as evidently would occur in a comet, then it is unlikely that this would lead to a complete dispersal of the dust particles imbedded within them. Many of the dust particles would stick together to form the filamentary networks and give low mean densities, such as are inferred and observed from meteors. It is also possible that some of the grain cores stick together as a result of collisions in interstellar space, during conditions in which the cores are largely stripped free of their mantles.

6. METEORITES

A typical stone meteorite consists of a number of more or less round stony inclusions imbedded in a matrix with a very fine-grained structure, representing an accumulation of small particles. These small particles are composed of silicates, oxides, and metals. The inclusions are called chondrules; most of them are of the order of a millimeter in radius, but some of them are significantly larger. Anders (1964) has postulated that the chondrules consist of very refractory materials, such as would condense only at high temperatures in a gas of solar composition, and that the matrix material is composed of compounds with a much wider range of volatility. This hypothesis has withstood a number of tests and generally appears to give quite a good description of the situation. These meteorites seem to have been assembled into their parent body at a temperature of about 450 K. Thus, certain of the more volatile elements (such as mercury, lead, thalium, bismuth, and indium) which can be expected to be in gaseous form at this temperature, are strongly depleted throughout. If the material is subjected to a considerable amount of heating while in its parent body, a metamorphism of the mineral phases tends to obliterate some of these sharply defined features.

The carbonaceous chondrites differ from the ordinary chondrites by containing water and a variety of complex organic compounds. The mass of these additional volatile elements ranges from a few percent in C3 meteorites to several tens of percent in C1 meteorites. C3 meteorites have a fairly large ratio of mass in chondrules to mass in matrix material; this ratio is considerably diminished in C2 meteorites, and C1 meteorites have no chondrules at all. In addition to ordinary chondrules, one C3 meteorite, Allende, contains a number of inclusions of extremely refractory calcium, aluminum, titanium oxides, and silicates. These are composed of the highest temperature condensates that can exist within the primitive solar nebula. There is a significant difference in the matrix composition of the carbonaceous chondrites; all the metals in them have been oxidized. Cosmothermometers indicate an assembly temperature of about 350 K for these meteorites.

The interpretation of these temperatures as assembly temperatures may be a bit artificial. What is really meant is that the chemical interaction between the condensed solids and the surrounding gas ceased at these temperatures. The assembly of the solids into larger bodies is one way of terminating this interaction. Another way would be to remove the gas from the vicinity of the particles. The latter seems to be a somewhat more likely interpretation in many cases, since many of the mineral grains in meteorites show radiation damage from exposure to ionizing particles, presumably from solar and galactic cosmic rays, which would not be able to penetrate through the gas in the primordial solar nebula. It seems more likely that the matrix material in such meteorites remained as individual particles in interplanetary space until the T Tauri-phase solar wind removed the gas of the primitive solar nebula from the region of the asteroid belt. Then these dust particles, together with chondrules, collected on the surfaces of asteroidal bodies, from which they have been later ejected by collisions.

It is reasonable to suppose that the matrix material has been derived from interstellar grains in a straightforward way. The scale of the fine structure in the matrix is comparable to the expected dimensions of interstellar grains and their cores. When interstellar grains, or clumps of such grains, are subjected to the solar-nebula conditions expected in the asteroid belt, then their mantles will be lost by evaporation. The mantles probably contain quite a number of more refractory elements imbedded within their matrix; these will probably be temporarily lost to the gas phase, but a recondensation onto the grain cores should follow shortly. Thus, the grains should quickly approach the chemical equilibrium conditions corresponding to the indicated "assembly" temperatures, and evidently they remain behind after the gas is removed by the T Tauri-phase solar wind, where they will be bombarded by solar cosmic rays and ultimately collected on asteroidal surfaces.

It is not entirely clear from this whether the iron should have been mostly in metallic form or in the form of magnetite in the interstellar grains from which the meteorites were derived. Under chemical equilibrium conditions, iron should be in metallic form above 400 K, and it should appear as magnetite below this temperature. Thus, the fact that the iron appears to be all oxidized in the carbonaceous chondrites could result either from its presence in this form in the interstellar grains or by chemical oxidation taking place through interaction of the grains with gas before the gas was removed from contact with the grains. Ordinary chondrites contain both metallic iron and magnetite, and since the materials in the matrix of these meteorites were subjected to temperatures above the limit at which magnetite can be reduced to metallic iron, one simple interpretation might be that the bulk of the iron was in magnetite form in the grains and that some of this was in sufficiently intimate contact with the gas to be reduced before assembly into the meteorites.

A related issue is the question of the remanent magnetism observed in many of these magnetite grains (Brecher, 1973). Since this appears to be predominantly a chemical remanent magnetism, the simplest interpretation is that the precipitation that caused the magnetite grain growth occurred in an environment where the local magnetic field lay in the range 0.1 to 1 G. However, it does not seem to exclude the possibility that the magnetism might be acquired as a chemical remanent magnetism in the precipitation of metallic iron, with the remanent magnetism preserved through the oxidation process that would convert the iron to magnetite. Since magnetic-field strengths of this order occur commonly in stellar-surface environments near which the grains may have been formed as cores before ejection into the interstellar medium, I suggested in Section 2 that the remanent magnetism associated with iron or magnetite may have been produced during the original nucleation process. However, the investigators in this field always interpret the remanent magnetism as being acquired through the production of the grain in a region of the primitive solar nebula that has a large magnetic-field strength. It seems to me very unlikely, however, that such large magnetic-field strengths should exist anywhere in the asteroidal-belt region of the primitive solar nebula, and furthermore it seems most unlikely that the nucleation and growth of magnetite grains could occur under such circumstances. If the grains accumulate gently enough onto the surfaces of asteroidal bodies, then those bodies need have only quite weak magnetic fields in order to produce an alignment of the magnetic moments of the incoming magnetite grains.

The chondrules in the meteorites present a number of interesting problems. These appear to have been formed as molten rock in space away from contact with other solid bodies. Furthermore, it seems that they must have cooled fairly rapidly through the freezing point. We would like to know if they have any relation to the interstellar grains.

Some time ago, Wood (1962) suggested that the chondrules were formed by condensation in a gas-pressure regime that would lead to the liquid phase of the chondrule materials. Unfortunately, this requires pressures very much higher than those expected to be present in the primitive solar nebula. Whipple (1966) suggested that lightning discharges in the primitive solar nebula would cause cylindrical implosion of the plasma column by the pinch effect; he postulated that this would cause small dust particles to be thrown together rapidly enough to produce melting. However, it does not appear likely that the density of the dust particles would be high enough for this to be effective. Cameron (1966) suggested that if dust grains had collected together into fluffy balls within the primitive solar nebula, then lightning flashes would produce enough hot electrons and ions in the vicinity of the plasma column to melt the dust balls into chondrules. However, it appears more likely that the surfaces would be vaporized and the interiors would remain unmelted. Both Whipple and Cameron suggest that chondrules may be formed as a result of collisions of dust balls: Whipple (1972, 1974, this volume) mentioning the possibility of electromagnetic discharges near the surfaces of larger bodies and Cameron (1973a) considering collisions with each other. However, the chondrules found in lunar materials, which presumably are produced by impact, differ greatly from chondrules in meteorites, through the inclusion of large amounts of un-melted material (Wood, 1974, this volume). Recently, it has seemed to be desirable to return to Wood's mechanism of producing chondrules in association with gas at high pressures, so that the chondrules pass through their liquid fields. While it should not be possible to do this anywhere in the original primitive solar nebula, it may be possible to do so later on after quite large planetary bodies have been accumulated within the primitive solar nebula. The solar-nebula gas becomes concentrated toward these bodies, producing the necessary higher pressures and temperatures therein. The solar nebula should flow past such bodies at a rather rapid rate, leading to wind-driven turbulence, which should extend to the level at which the chondrules will exist within their liquid states, and which may be able to remove the chondrules fairly rapidly to the other parts of the undisturbed primitive solar nebula.

The last suggestion would remove most traceable connections between the original interstellar grains and the final chondrules. Fluffy dust balls can be convected toward higher temperatures and pressures, where they can be melted, and then convected away fairly rapidly. On the other hand, an element of the gas rising from a level at still higher temperatures, at which everything is vaporized, to a level at lower tem-perature may have an independent nucleation of condensable material, forming chon-drules and refractory inclusions. It is interesting to note that the condensation products arising in this way may become quite large, since this process is analogous to the slow expansion of a cloud chamber, in which only a few droplets of water may be formed. The other situation, in which a cloud chamber is rapidly expanded, leading to a large number of small droplets, is more analogous to the case of expulsion of gas from a stellar atmosphere, in which the condensation products should be very small. But there should be few distinguishing features between raising a fluffy dust ball to the melting point and forming a new condensation.

7. COSMOCHRONOLOGY

It is necessary to reexamine the basis for cosmochronology in the light of the picture presented in this paper. Meteoritic cosmochronometers depend on the radioactivities contained within the meteorites. Two situations arise: cases in which the daughter product of the radioactivity is a gas, and those in which it is a solid.

Let us first consider the gas cases. The two important gas cosmochronometers are ^{129}I and ^{244}Pu. These radioactivities have half-lives of 17 and 81 million years, respectively. Hence, they are now extinct in the solar system, but they were present in the original material out of which the solar system was formed. One can estimate the efficiency with which either of these substances was retained in the parent material of any sample by measuring the abundance of ^{127}I, a stable isotope, and the abundance of uranium, which has chemical properties similar to that of plutonium. Both these radioactivities produce stable isotopes of xenon, a rare gas, the iodine radioactivity by β decay, and the plutonium radioactivity by spontaneous fission. A time interval, called a "formation interval," is measured for each of these radioactivities by determining the amount of their xenon daughter products in a meteoritic sample and, hence, deducing the amount of the parent substance present in the sample at the time it became capable of retaining the gas products against diffusion losses. This deduced parent abundance is then compared with the expected level of abundance of the radioactivity in the interstellar medium, as it would be maintained there by continuing stellar nucleosynthesis. Such formation intervals are clearly model-dependent, since the initial level of radioactivity in the interstellar medium must be estimated on the basis of a model of the nucleosynthesis contributions. If one assumes that the recent rate of production of these radioactivities by nucleosynthesis has been about half the mean rate over the past history of the Galaxy, then formation intervals of about 2×10^8 years are determined.

This number is a measure of the time between the last major nucleosynthesis episode affecting the material that formed the solar system and the epoch at which the system had formed and the meteoritic parent bodies were cool enough to retain xenon against diffusive losses. I shall not go into the details of this interpretation here. However, it is clear that this cosmochronometer would give very different results if the radioactive iodine and plutonium had become parts of the cores of the interstellar grains. In such a case, it is likely that much of the xenon decay from these radioactivities would have been retained in the grain cores, and thus introduced into the meteorite parent bodies, if the temperatures to which the material was subjected did not exceed about 450 K. However, if the iodine and plutonium were incorporated only in the grain mantles, then xenon decay products would be lost into the gas of the primitive solar nebula and the chronometers would operate as usually interpreted.

It is thus useful to note that both these radioactivities are products of the r-process, and are produced in supernova explosions. We noted above that the products of supernova explosions should probably not participate in the grain-nucleation process, but should be ejected directly into interstellar space. This indicates that the iodine and plutonium radioactivities should be picked up by the interstellar grains only in their mantles, and hence the conventional interpretations of the formation intervals should be satisfactory.

There are two principal cosmochronometers that produce solid products. These involve the decay of ^{87}Rb to ^{87}Sr and of ^{235}U and ^{238}U to ^{207}Pb and ^{206}Pb, respectively. Measurements consist of the determination of the variations in isotopic abundances in strontium and lead. Here the radioactive half-lives are longer, and the measured amount of radioactive decay is usually used to determine the time since the last chemical differentiation of the sample. Underlying this method is the basic assumption that, at the start of the time interval in question, one knows the isotopic composition of the daughter substance that forms an accumulated background in the measured sample.

^{87}Rb and the uranium isotopes are also products of the r-process of nucleosynthesis. Thus, these radioactivities should also become part of grain mantles in the interstellar medium. As such, they should also be lost to the gas phase when the grains are brought into the asteroidal-belt region of the primitive solar nebula, but they should also be rapidly reaccumulated onto the solid material. Most of the daughter-product strontium should be similarly reaccumulated, and although much of the lead daughter product may not be reaccumulated, the actual abundance of this material is not relevant to the chronology measurement – but only the isotopic composition of the material is. Hence, the conventional basis for the interpretation of these cosmochronometers should also be preserved.

So far, there seems to be no difficulty with cosmochronometry, but some results in this field appear at first sight to be inconsistent with the general picture of solar-system formation that I have outlined. Variations exist between different meteoritic samples in the formation interval; also, variations of a few tens of millions of years occur in the rubidium–strontium and lead–lead ages of some of the meteorites. These differences are several orders of magnitude greater than the time interval in which I expect the solid bodies of the solar system to be assembled in the presence of the primitive solar nebula; this is where the apparent inconsistency arises.

However, I have recently suggested, first, that the abundances of r-process radio-activities should be inhomogeneously distributed in interstellar space owing to the difficulty of the diffusion of nucleosynthesis products across the lines of force of the inter-stellar magnetic field, and, second, that a small part of these differences may be preserved in local patches within the primitive solar nebula at the time of assembly of the meteorite parent bodies (Cameron 1973b). These considerations predict that there should also be very small differences in the abundances of the stable isotopes of the daughter elements.

A preliminary attempt to determine whether such differences exist in rubidium and strontium has been made by Gray, Papanastassiou, and Wasserburg (1973). Their errors in the relative abundances of the rubidium isotopes amount to an order of magnitude greater than the expected small differences, and both the errors and the variations in the measured values for the strontium isotopes are comparable to the expected differences. Therefore, this remains an unsettled question that I hope can be resolved by further measurements, possibly on other elements.

8.   RECYCLING OF SOLAR-NEBULA SOLIDS

The solar-nebula models presented in Figures 1 to 3 contain far more chemically condensable material than now appears in the planets in the solar system. Throughout the inner regions of the solar nebula, the bulk of these condensed solids will be trans-ported inward with the gas that forms the sun. However, the gas in the outer part of the solar nebula does not become a part of the sun, but remains at fairly large distances in order to take up the angular momentum transported outward from the center. Thus, the solid materials that should exist at that distance will be left within the solar system at the time the gas is removed by the T Tauri-phase solar wind. Since this material should be considerably more massive than the planets Uranus and Neptune, it is interesting to ask what became of it. If it were to spiral inward as a result of the Poynting-Robertson effect, it would be surprising if a large amount of it were not collected by the giant planets.

As I have discussed in more detail elsewhere, probably the majority of this material is subject to destruction by rotational bursting. In this process, irregularities in the absorption or scattering of sunlight can be expected to result in small torques, which will be applied to small bodies as they rotate. These torques lead to angular acceleration of the bodies, and the torques will continue as the bodies spin faster and faster. Ultimately, the bodies will be spinning so rapidly that the internal cohesion will become inadequate to maintain the bodies against rotational bursting (Radzievskii, 1954; Paddack, 1969). The fragments, in turn, should themselves be subject to angular acceleration and further destruction by rotational bursting. Since these bodies are likely to consist of fragile filamentary frameworks of adhering interstellar grain cores, it is likely that the process of subdivision by rotational bursting can continue until the bodies have been reduced to their constituent grain cores. These can then be ejected from the solar system by radiation pressure.

Thus, much of the leftover finely divided mass within the solar system should be returned to the interstellar medium in the form from which it was originally obtained: the cores of interstellar grains. Material from farther out in the solar system should probably be returned also with the original grain mantles, perhaps slightly simplified chemically by mild heating. Since these mantles are likely to be lost in the course of a few million years anyway, by photosputtering, this should make very little difference to the ultimate result. However, all the bodies larger in size than about an orange should be too massive for these torques to be very effective in angular acceleration, and also too large for significant inward spiraling by the Poynting-Robertson effect; such bodies may still be members of the solar system. It is very difficult to know whether there has been an ultimate significant depletion of the gases returned to the interstellar medium from the primitive solar nebula in their condensable solids as a result of the formation of these bodies, but there has undoubtedly been a large depletion of condensable materials from the gas in the satellite nebulae forming the comets.

Thus, I differ with Herbig's (1970) view that the solar nebula can be a significant source of interstellar grains. It appears that interstellar grains can be recycled through the solar nebula to a significant degree, but I would expect that any new chemical condensations resulting from original nucleation within the solar nebula would be of a sufficiently large size that radiation pressure would be unable to eject them from the solar system. This is again the cloud-chamber effect: Condensation occurring in a slowly cooling gas or in a gas that expands very slowly should produce much larger particles than would condensation in a gas more rapidly ejected from a stellar atmosphere.

9. SUMMARY

The principal point I have tried to make in this paper is that there should be an underlying unity between the character of interstellar dust and interplanetary dust. The cores of interstellar grains should be mainly formed when mass loss takes place from stellar surfaces. Fragile volatile mantles are added when these grain cores reside in interstellar space. The mantles will be periodically removed from the grain cores, depending on the character of the local interstellar conditions. When star formation occurs, with the accompanying formation of a planetary system, the interstellar grains will not be completely evaporated in those regions of the primitive solar or planetary nebula in which planets are formed. The farther out one goes in the solar nebula, the greater the content of the original interstellar grains that are retained. Comets should be formed essentially out of entire interstellar grains, with at best some chemical simplification in the grain mantles. Meteorites should be formed from interstellar grain cores, from which the volatile mantles have been evaporated. Interplanetary dust should consist of fragile filamentary clumpings of interstellar grain cores. Many of these interstellar grain cores will be returned to the interstellar medium toward the end of the formation period of a planetary system.

Interstellar grains and interplanetary grains have been studied as separate disciplines. It is my belief that these topics have a great deal to teach each other as a result of this underlying unity in their subject material.

ACKNOWLEDGMENTS

I am indebted to E. Anders, A. Brecher, and J. A. Wood for help in guiding my thinking during the development of these ideas. However, they should not be charged with any responsibility for the form these conclusions have finally taken. This research has been supported in part by grants from the National Science Foundation and the National Aeronautics and Space Administration. It is a pleasure to dedicate this work to Fred Whipple on the occasion of his retirement as the Director of the Smithsonian Astrophysical Observatory.

REFERENCES

Aannestad, P. A., 1973. Astrophys. Journ. Suppl. No. 217.

Aannestad, P. A., and Purcell, E. M., 1973. Ann. Rev. Astron. Astrophys. 11, 309.

Anders, E., 1964. Space Sci. Rev. 3, 583.

Anders, E., 1972. In From Plasma to Planet, ed. by A. Elvius (Almqvist and Wiksell, Stockholm).

Blander, M., and Katz, J. L., 1967. Geochim. Cosmochim. Acta 31, 1025.

Brecher, A., 1973. In Evolutionary and Physical Properties of Meteoroids, ed. by C. L. Hemenway, P. M. Millman, and A. F. Cook, NASA SP-319, Washington, D.C.

Cameron, A. G. W., 1966. Earth Planet. Sci. Lett. 1, 93.

Cameron, A. G. W., 1971. Nature 229, 178.

Cameron, A. G. W., 1973a. Icarus 18, 407.

Cameron, A. G. W., 1973b. Nature 246, 30.

Cameron, A. G. W., and Pine, M. R., 1973. Icarus 18, 377.

Geisel, S. L., 1970. Astrophys. Journ. (Lett.) 161, L105.

Geisel, S. L., Kleinmann, D. E., and Low, F. J., 1970. Astrophys. Journ. (Lett.) 161, L101.

Gilman, R. C., 1969. Astrophys. Journ. (Lett.) 155, L185.

Goldreich, P., and Ward, W. R., 1973. Astrophys. Journ. 183, 1051.

Gray, C. M., Papanastassiou, D. A., and Wasserburg, G. J., 1973. Icarus 20, 213.

Grossman, L., 1972. Geochim. Cosmochim. Acta 36, 597.

Herbig, G. H., 1970. Mem. Soc. Roy. Sci. Liège, Ser. 5 19, 13.

Huffmann, D. R., 1973. In Interstellar Dust and Related Topics. Proc. IAU Symp. No. 52, ed. by J. M. Greenberg and H. C. van de Hulst (D. Reidel Publ. Co., Dordrecht, Holland).

Mészáros, P., 1972. Astrophys. Journ. 177, 79.

Mészáros, P., 1973a. Astrophys. Journ. 180, 381.

Mészáros, P., 1973b. Astrophys. Journ. 180, 397.

O'Keefe, J. A., 1939. Astrophys. Journ. 90, 294.

Oort, J. H., 1950. Bull. Astron. Inst. Netherlands 11, 91.

Öpik, E. J., 1973. Astrophys. Space Sci. 21, 307.

Paddack, S. J., 1969. Journ. Geophys. Res. 74, 4379.

Parker, E. N., 1966. Astrophys. Journ. 145, 811.

Perri, F., and Cameron, A. G. W., 1974. Icarus, in press.

Podolak, M., and Cameron, A. G. W., 1974. Icarus, in press.

Radzievskii, V. V., 1954. Dokl. Akad. Nauk SSSR 97, 49.

Roberts, W. W., 1969. Astrophys. Journ. 158, 123.

Roberts, W. W., and Yuan, C., 1970. Astrophys. Journ. 161, 877.

Sparks, W. M., and Kutter, G. S., 1974. Studies of hydrodynamic events in stellar evolution. III. Ejection of planetary nebulae, preprint.

Watson, W. D., and Salpeter, E. E., 1972. Astrophys. Journ. 174, 321.

Whipple, F. L., 1948. Harvard Obs. Monograph No. 7, 109.

Whipple, F. L., 1950. Astrophys. Journ. 111, 375.

Whipple, F. L., 1966. Science 153, 54.

Whipple, F. L., 1972. In From Plasma to Planet, ed. by A. Elvius (Almqvist and Wiksell, Stockholm).

Wilson, R. E., 1971. Astrophys. Journ. 170, 529.

Wood, J. A., 1962. Geochim. Cosmochim. Acta 26, 739.

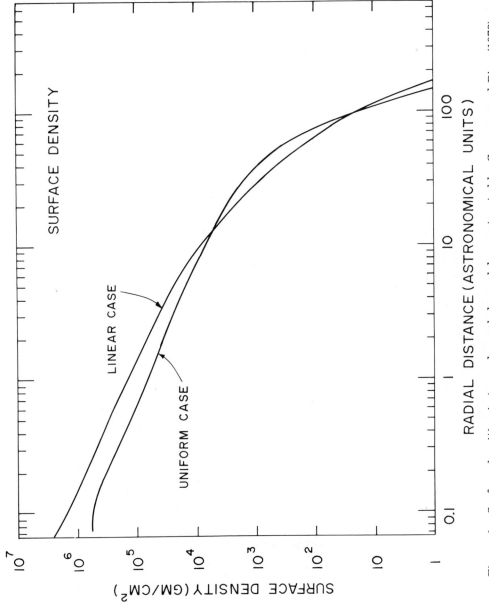

Figure 1. Surface densities in two solar–nebula models constructed by Cameron and Pine (1973).

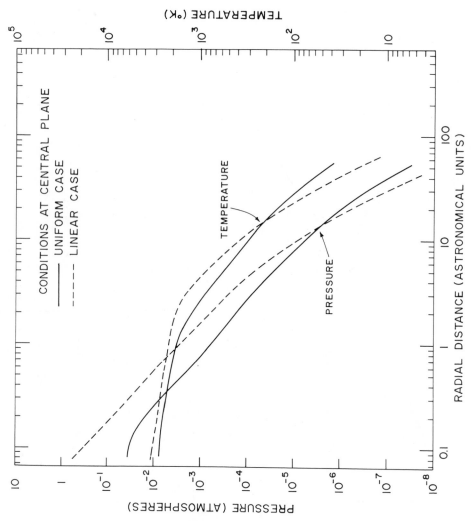

Figure 2. Temperatures and pressures in the Cameron–Pine solar–nebula models.

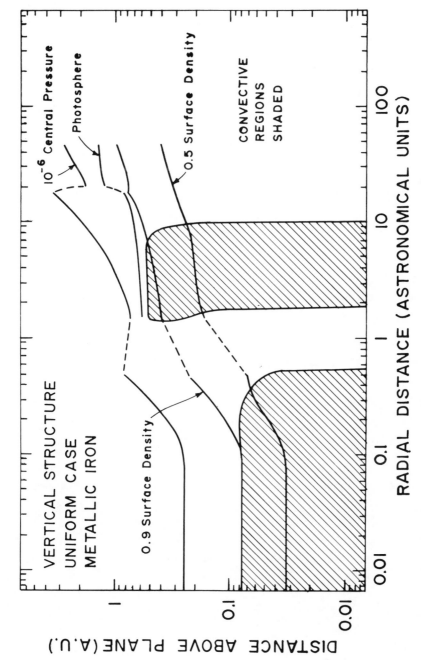

Figure 3.   The vertical structure in one of the Cameron–Pine solar–nebula models.   Lines indicate levels at which 0.5 and 0.9 of the surface density lie toward the central plane, the level at which the central pressure has fallen by a factor of 10^6, and the photospheric level from which radiation is emitted. The shaded regions contain convective gas motions.

CHEMISTRY OF SOLAR MATERIAL

Stephen S. Barshay and John S. Lewis

Massachusetts Institute of Technology, Cambridge, Massachusetts

ABSTRACT

The calculated chemical compositions of the gaseous and condensed phases in the primitive solar nebula are presented for both equilibrium and disequilibrium condensation. The implications for the compositions of individual solar-system bodies will be briefly discussed. Condensation from an otherwise solar-composition gas in which carbon is more abundant than oxygen is mentioned.

1. INTRODUCTION

Those current models for the origin and evolution of the solar system that bear
some resemblance to the observable properties of planets and meteorites usually
share the proposition that the solid material from which the planets formed originated
in a homogeneous, solar-composition cloud of gas and dust called the primitive solar
nebula (Cameron and Pine, 1973; Cameron, 1973a). The purpose of this paper is
to summarize the chemical reactions that may occur in such a solar nebula, and to
describe the chemical composition of the condensates that form from the gas. The
implications of this theory to the composition of individual solar-system bodies will be
briefly discussed.

In a discussion of condensation from a gas, we consider two extreme types of
condensation processes: equilibrium condensation and disequilibrium condensation.
It is important to remember, however, that condensation can occur with characteristics
intermediate between these two extremes. In the equilibrium case, it is assumed that
conditions in the primitive solar nebula changed so slowly that the gaseous and condensed phases were always in thermodynamic equilibrium. The composition of the
condensed phase was, in the language of thermodynamics, a "state function," dependent
only on the local temperature, pressure, density, and elemental composition of the
solar nebula at the moment when condensation processes ceased. The past history of
the nebula was of no consequence; whether it was warming or cooling, expanding or
contracting, condensing or vaporizing, eating or sleeping made no difference in the
final composition.

In the case of disequilibrium condensation, the condensed phase is removed from
interaction with the gas phase as quickly as it is formed. No chemical reactions between
gas and solid, or even between two solid phases, is allowed. The mechanism of such a
process would be the rapid accretion of condensed material into larger bodies, effectively isolating that material from reaction with the gas phase and creating layered or
onion-skinned planets, as first suggested by Eucken (1944).

In this discussion, we will ignore the question of how the initial condensate formed. The accumulation of the first 50 or so gas molecules into a crystal is a thermodynamically unfavorable process, which accounts for the phenomenon of supersaturation of gases and liquids. However, once these initial crystals are large enough to be stable, they will act as nucleating centers for further condensation. Perhaps cooling in the primitive solar nebula occurred so very, very slowly that random nucleation could occur without supersaturation, or the original dust grains, which existed before the nebula began to contract, were never fully vaporized, or possibly supersaturation did occur during condensation. These and other possible scenarios are still open to speculation and research.

2. THE INNER SOLAR SYSTEM

The outstanding features of equilibrium condensation in the primitive solar nebula are shown in Figure 1. Above the curve labeled W, all species are stable only in the gas phase. The W condensation curve represents the condensation temperature of a whole class of refractory, or high-temperature, metals, including Os and Ir (see Grossman, 1972a). $CaTiO_3$ represents a class of refractory oxides, including compounds of Al, Ca, Ti, V, Zr, the rare earths, U, and Th. Next are the condensation curves of Fe–Ni alloy and solid $MgSiO_3$, which together dominate the chemistry of the terrestrial planets. The triangular region below the Fe curve at the upper right is the stability field of molten iron. Grossman (1972b) has calculated that, at temperatures 100 to 200 K below the $MgSiO_3$ curve, the alkali aluminosilicates become stable, including compounds of K, which is extremely important as an internal heat source for the planets. The temperature at which H_2S gas reacts with Fe metal to form FeS is noted, as is the temperature at which the last of the remaining metallic iron is oxidized by water vapor to FeO. The appearance temperatures of tremolite (a hydrous calcium silicate), serpentine (a hydrous ferromagnesium silicate), and ice complete the diagram.

Each planet will initially contain those condensates that appear at or above its formation temperature. Thus, Mercury will consist of the refractory metals and minerals, iron–nickel alloy, and a limited amount of $MgSiO_3$. This accounts for the observed density of Mercury, which is too low for a body made mostly of metallic iron and too high for a body made of iron plus enstatite in solar proportions (Lewis, 1972a).

Venus, in addition to everything mentioned above for Mercury, contains a full solar complement of enstatite plus quantities of alkali aluminosilicates, but no sulfur or water.

Fortunately for us, the earth accreted in a region where tromolite, the most stable of the water-bearing silicates, was at least marginally stable. Iron is present in the earth in three different forms, as Fe metal, FeO, and FeS. Metallic iron and FeS form a low-melting eutectic mixture that was probably the first material to melt as the interior of the earth heated up, leading to autocatalytic differentiation and the formation of a conducting core very early in the earth's history (Murthy and Hall, 1970).

Mars should contain even more water than the earth, since it has a full complement of tremolite, and additional quantities of serpentine, another hydrous silicate. It seems likely that much of this water is still bound up in the rocks of Mars, awaiting a time when temperatures will be high enough to volatilize it. The iron on Mars has all been oxidized to FeS or FeO, so that a metallic-iron core is unlikely.

The implications of the equilibrium-condensation sequence for the fraction of available mass condensed and for the cumulative density of the condensate are shown in Figure 2 and Table 1. The beautiful way in which the model matches the observed densities of the terrestrial planets gives some justification for believing that condensation and accretion in the inner solar system occurred under equilibrium conditions.

An element-by-element review of the equilibrium chemistry of solar material is provided in Figure 3. The fate of iron during equilibrium condensation will illustrate the use of the chart. Metallic iron—nickel alloy condenses at ~1500 K, oxidation by H_2S gas to form FeS (troilite) at ~680 K results in the exhaustion of available sulfur, but leaves about half the iron unreacted until ~500 K when the last remaining iron metal is oxidized to FeO, which is incorporated into olivine. Olivine, in its turn, reacts with more water vapor at ~300 K to make the hydrous mineral serpentine. Our present knowledge of the solar system, including the observed carbon content of comet nuclei and the methane observed in the atmospheres of outer solar-system bodies, does not appear to require condensation temperatures much below 50 K if processes of equilibrium or near-equilibrium condensation are assumed. This is illustrated in the figure by a dashed horizontal line.

3. THE OUTER SOLAR SYSTEM

As one moves radially outward from the center of the primitive solar nebula, the temperature and pressure decrease and conditions become acceptable for the condensation of more and more material. Regardless of whether we assume equilibrium condensation, as described in Figure 2, or disequilibrium condensation, as described in Figure 4, we see from Cameron's adiabat (Figure 1) that the asteroid belt should be populated with rocky material devoid of ices. The measured densities of some asteroids, and the recent spectroscopic observations of Chapman, McCord, and Johnson (1973), support this conclusion.

Beyond the asteroid belt, we enter the region where ices are stable, and we expect the planets and satellites of the outer solar system to be an admixture of rock-forming and ice-forming elements. The observed densities of some of the satellites of Jupiter, Saturn, and Neptune support this hypothesis.

Details of the composition of the icy material depend on whether condensation was fast or slow, as illustrated in Figure 5. In the disequilibrium case, a condensate, once formed, is not permitted to react with anything else. This is not true in the equilibrium case where, for example, ammonia and methane gases react with the previously condensed water ice. The formation of the ammonia solid hydrate exhausts the available ammonia, and methane clathrate-hydrate formation uses up all the remaining water ice, leaving most of the methane still in the gas phase to condense as methane ice at even lower temperatures. Thus, equilibrium or near-equilibrium conditions permit an object to retain ammonia and methane at higher temperatures than are possible under disequilibrium conditions. Also, sulfur is quantitatively removed from the gas as FeS at 680 K if the condensed iron metal is allowed to equilibrate with the gas; otherwise, sulfur does not condense in appreciable quantities until the formation of $NH_4SH(s)$ at a much lower temperature.

Cameron's (1973b) discussion of the origin and history of the outer planets proposes that they formed when sections of the solar nebula became gravitationally unstable and collapsed, much as the sun formed from the material at the center of the nebula. Alternatively, Cameron suggests that the Jovian planets may have formed from a

condensed rock-and-ice core, which, due to its great mass, gravitationally captured
large amounts of otherwise uncondensed gases, including hydrogen, helium, methane,
and ammonia. Either of these processes would result in planets composed of
unfractionated solar material, that is, all the elements in their solar proportions.

Weidenschilling and Lewis (1973) and Lewis (1972b) have discussed the composition of the cloud layers of the Jovian planets in terms of a disequilibrium-condensa-
tion model. They predict, for example, the formation of an ammonia-ice cloud
layer at the level in the Jovian atmosphere where the temperature is about 130 K.
Above that level, where the temperature is lower, any condensed ammonia will fall
toward the cloud layer. Below the cloud, the temperature will be higher than 130 K,
and any condensed ammonia will be vaporized. In this way, individual cloud layers
of ammonia-water solution, water ice, $NH_4SH(s)$, $NH_3(s)$, $CH_4(s)$, and $Ar(s)$ are
possible on various planets.

4. CARBON AND CHONDRITES

Hydrogen dominated the chemistry of the gas phase in the primitive solar nebula,
which is not surprising since 93% of the atoms were hydrogen. H_2, CH_4, NH_3,
H_2O, and H_2S were the most abundant gases of their respective elements at low temperatures. At higher temperatures, however, these compounds become thermally
unstable. Methane, for example, is oxidized above ~800 K by water vapor to give H_2
and CO, both of which persist beyond 3000 K.

Carbonaceous chondrites, which contain elemental carbon and primitive prebiotic
matter, are a remarkable puzzle, since nowhere in the chemistry of solar material do
we see any condensed carbon other than methane ices. Anders, Hayatsu, and Studier
(1973) have proposed that the hydrocarbons are the result of incomplete reduction of
carbon monoxide by hydrogen in the presence of magnetite and silicate catalysts. But
this does not explain the origin of graphite.

These and other questions have led to recent interest in our laboratories in the
chemistry of carbon-rich material, that is, a system with much the same relative
elemental abundances as our solar system, except that there is more carbon than

oxygen. Knowledge of the chemistry of such a system may be applicable to the carbonaceous chondrites and to the atmospheres of carbon-rich stars.

The presence of more carbon than oxygen in the nebular gas means that practically all the oxygen will be in the form of carbon monoxide, with nothing left over for formation of water vapor. The condensation of refractory oxides, magnesium silicates, and alkali aluminosilicates is impossible under these conditions. In fact, it is unlikely that any oxygen-containing compound will condense at temperatures above ~600 K, at which point the CO is reduced to CH_4 and H_2O. Instead, graphite is expected as a high-temperature condensate, with various carbides, sulfides, and nitrides condensing at lower temperatures. These preliminary conclusions are, where applicable, in agreement with work done by Gilman (1969), and Tsuji (1964).

Lewis (1973a) has pointed out that the presence of FeS, FeO-bearing silicates, and hydrous silicates in many meteorites strongly suggests some sort of chemical equilibration between gas and solid at low temperatures, even though complete equilibrium is often not achieved. The presence of magnetite (Fe_3O_4) in carbonaceous chondrites is an interesting exception to the condensation sequences we have discussed. Its formation appears to require interaction between water vapor and metallic iron to produce FeO but incomplete solid-solid equilibrium between the resultant FeO and magnesium silicates, so that magnetite is produced instead of olivine.

5. SUMMARY

We have seen that the observed densities of the terrestrial planets support a nearly equilibrium type of condensation process; in the outer solar system, however, the densities and properties of satellites and other objects are not known with sufficient accuracy to differentiate between the two processes. The cloud layers of the giant planets are proposed to be the product of inhomogeneous condensation processes in their atmospheres. Meteorites with low formation temperatures and carbon stars remain very incompletely understood.

ACKNOWLEDGMENTS

We are pleased to acknowledge support of the writing of this review and of much of the work described herein by the National Aeronautics and Space Administration under Grant No. NGL 22-009-521.

REFERENCES

Anders, E., Hayatsu, R., and Studier, M. H., 1973. Science 182, 781.

Cameron, A. G. W., 1968. In Origin and Distribution of the Elements, ed. by L. H. Ahrens (Pergamon Press, London) p. 125.

Cameron, A. G. W., 1973a. Icarus 18, 407.

Cameron, A. G. W., 1973b. Space Sci. Rev. 14, 383.

Cameron, A. G. W., and Pine, M. R., 1973. Icarus 18, 377.

Chapman, C. R., McCord, T. B., and Johnson, T. V., 1973. Astron. Journ. 78, 126.

Eucken, A., 1944. Nachr. Akad. Wiss. Göttingen, Math.-Phys. Kl., Heft 1, 1.

Gilman, R. C., 1969. Astrophys. Journ. (Lett.) 155, L185.

Grossman, L., 1972a. Ph.D. thesis, Yale University.

Grossman, L., 1972b. Geochim. Cosmochim. Acta 36, 597.

Lewis, J. S., 1972a. Earth Planet. Sci. Lett. 15, 286.

Lewis, J. S., 1972b. Icarus 16, 241.

Lewis, J. S., 1973. Space Sci. Rev. 14, 401.

Murthy, V. R., and Hall, H. T., 1970. Phys. Earth Planet. Interiors 2, 276.

Tsuji, T., 1964. Ann. Tokyo Astron. Obs. 9, No. 1.

Weidenschilling, S. J., and Lewis, J. S., 1973. Icarus 20, 465.

TABLE 1.

Order of reactions in Figure 2.

1	Condensation of Ca, Al, and Ti silicates
2	Metallic-iron condensation
3	Condensation of $MgSiO_3$, which drops the cumulative density below 6 g cm^{-3}
4	Conversion of $H_2S(g)$ to $FeS(s)$
*5	Reduction of CO to CH_4 and H_2O
6	Conversion of the last of the remaining iron metal to FeO
7	Conversion of enstatite to serpentine
*8	Reduction of N_2 to NH_3
9	Condensation of water ice
10	Condensation of NH_3 as solid $NH_3 \cdot H_2O$
11	Conversion of water ice to solid $CH_4 \cdot 8H_2O$
12	Condensation of the remaining methane as methane ice
13	Condensation of Ar(s)
14	Condensation of Ne(s)
15	Condensation of $H_2(s)$
16	Condensation of He(s)

*Gas-phase reaction.

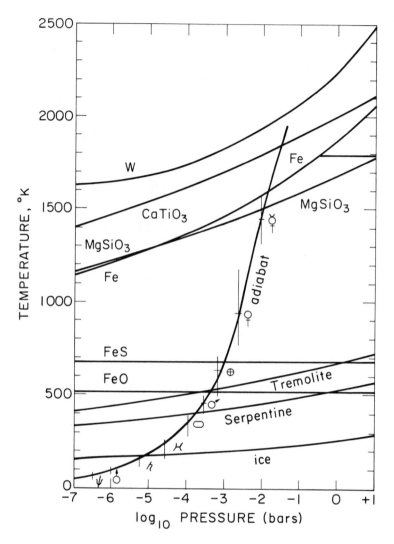

Figure 1.   Stability limits of condensates under equilibrium conditions in the solar
nebula, 0 to 2500 K and 10^{-7} to 10 bars total pressure.   The assumed
abundances of the major elements are discussed in Lewis (1972b); the
solar abundance of tungsten was taken from Cameron (1968).   An adiabat
is drawn that represents the present best estimate of the pressure and
temperature profile in the solar nebula, as developed by Cameron (1973a)
and Cameron and Pine (1973).   The regions corresponding to the formation conditions of the planets in Cameron's models are marked on the
adiabat.

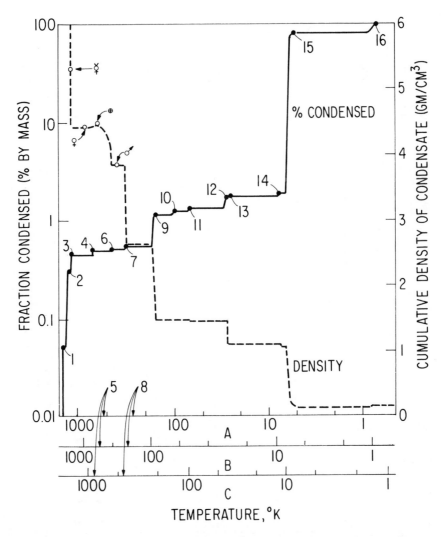

Figure 2.   Cumulative mass and density for the equilibrium condensation sequence at
various total pressures.   Approximate temperature scales are given for
pressure of (A) 10^{-6} bar, (B) 10^{-4} bar, and (C) 10^{-2} bar.   The zero-
pressure densities of the terrestrial planets are indicated on the cumulative
density curve (Lewis, 1972a).   The 16 steps indicated are listed in Table 1.
Steps 4 and 6 are pressure-independent, and should be read only on scale
B.   At pressures below 10^{-5} bar, step 2 occurs at a lower temperature
than step 3.   An Fe:Si ratio of 1.06 was used in computing the density of
the condensate, as discussed in Lewis (1972a).

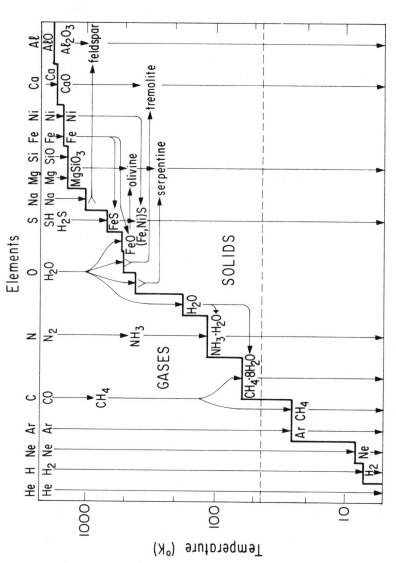

Figure 3.   Sequence of major reactions during slow cooling of solar material. This figure may be read as a flow chart showing the fate of each of the 15 most abundant elements during cooling from 2000 to 5 K at a total pressure of about 10^{-3} bar. The elements are listed across the top, and directly underneath are the major gas species of each element at 2000 K. The staircase separates gases from condensed phases. The dashed line represents the probable lowest temperature present in the primitive solar nebula during condensation. For details of the high-temperature reactions, see Grossman (1972b) and for low-temperature reactions, see Lewis (1972b).

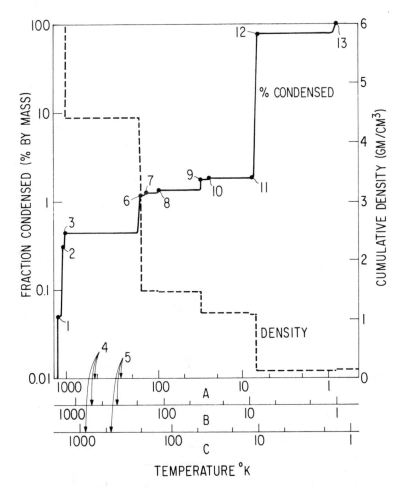

Figure 4. Cumulative mass and density for the disequilibrium condensation sequence at various total pressures. Approximate temperature scales are given for pressures of (A) 10^{-6} bar, (B) 10^{-4} bar, and (C) 10^{-2} bar. The 13 steps indicated are (1) condensation of Ca, Al, and Ti silicates, (2) metallic iron condensation, (3) condensation of $MgSiO_3$, which brings the cumulative density below 6 g cm^{-3}, (4) reduction of CO to CH_4 and H_2O, (5) reduction of N_2 to NH_3, (6) condensation of water ice, (7) condensation of solid NH_4SH, (8) condensation of ammonia ice, (9) condensation of methane ice, (10) condensation of Ar(s), (11) condensation of Ne(s), (12) condensation of H_2(s), (13) condensation of He(s). At pressures below 10^{-5} bar, step 2 occurs at a lower temperature than step 3. Steps 4 and 5 are gas-phase reactions, not condensations, and are indicated only on the temperature scales.

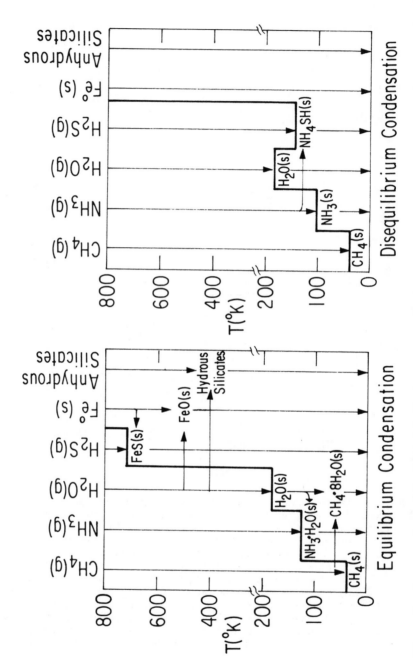

Figure 5.   A comparison of the fate of C, N, O, S, Fe, and silicates during equilibrium and disequilibrium condensation at low temperatures. The staircase separates gases from condensed phases, as in Figure 3. For further details, see Lewis (1973).

DUST IN STELLAR ATMOSPHERES

E. E. Salpeter

Cornell University, Ithaca, New York

ABSTRACT

Nucleation theory, developed for liquid-droplet growth, is reviewed. Difficulties in the growth of crystals held together by valence forces are pointed out. The rate of mass loss is estimated for dust grains in cool stellar atmospheres as a function of luminosity. For intermediate values of the luminosity, only a few high-ejection-velocity dust grains are ejected from the atmosphere.

1. INTRODUCTION

The comprehensive review of observational data by Woolf (1974, this volume) gives me a good excuse to restrict myself to purely theoretical considerations. Basic ideas on dust-grain production in cool stellar atmospheres were formulated by Hoyle and Wickramsinghe (1962), Donn and Stecher (1965), and others some years ago: If the surface temperature T_e of a giant star is sufficiently low, then dust grains will condense out; if the star's luminosity L is sufficiently large, the grains will be driven out into the interstellar medium by radiation pressure. Some numerical calculations (e.g., Donn et al., 1968; Fix, 1969) are already available, but I will describe only qualitative order-of-magnitude relationships; a full calculation would have so many interlocking (and at present uncertain) facets that it is useful to consider first the importance of various parameters.

I will cover two topics. The first, discussed in Section 2, is the condensation process in a gas undergoing cooling, where thermochemistry should reveal which minerals will condense out first and nucleation theory should give the size and number of seed nuclei and the grain size to which they eventually grow. Section 3 deals with the possible mass-flow rates and ejection velocities at which grains, produced in a stellar atmosphere, are expelled and drag some gas with them. I will consider only stars that are cool enough for grains to form right in the photosphere, but where there would be no mass loss without grains. Novae are an example of a different class of stars, which have a photosphere too hot for grain formation but have a mechanism (independent of grains) for mass ejection. As Woolf shows, this class is quite important observationally, but I will disregard it simply because the theory for it is even less well understood than that for cool stars.

2. NUCLEATION AND GRAIN GROWTH

Before reviewing nucleation theory, I want to summarize the thermochemical data, which specify the minerals that would condense out as a function of pressure and temperature under conditions of thermal equilibrium. Similar considerations for the

early solar-system disk are reviewed by Lewis (1974, this volume), but for
our stellar atmospheres, we need temperatures 2 or 3 times higher, $T \sim 1000$ to 2000 K,
and somewhat lower pressures. At these higher temperatures and lower densities, two
features of the chemistry of carbon and oxygen appear, illustrated schematically in
Figure 1. First of all, because of the stability of carbon monoxide gas, whichever of
carbon and oxygen has the lower abundance (by number) is completely fixed in the form
of CO. For this reason, we need to consider either graphite-like grains rich in
carbon or oxygen-rich minerals (such as oxides and silicates), but not both. Another
high-temperature, low-density feature of the carbon-rich cases is that the gas-phase
precursor molecules are not methane, but acetylene (C_2H_2) at medium densities and
C_2 or C_3 at very low densities.

In the expansion phase of a long-period variable star (for instance, a Mira
variable) or in the updraft of a convection cell — or simply in a steady outward stream
in a stellar atmosphere — all the gas is undergoing cooling. Some molecular species
become supercooled first and a nucleation process begins. Numerical calculations
undertaken so far have relied on a "classical nucleation theory," primarily developed
for terrestrial experiments, where gas densities are higher than in stellar atmospheres
and where liquid droplets are formed that bind molecules weakly compared with
the valence-force binding in graphite, oxides, or silicate crystals. Some of the finer
details of classical nucleation theory are therefore not strictly applicable. For
instance, because of difficulties with transfer of the deposited latent heat (Salpeter,
1973), nuclei cannot be grown solely by the deposition of monomers from the gas phase
one by one; in some instances, a diatomic or polyatomic molecule impinging from the
gas phase with the breaking of a bond is required, one part of the molecule bonding to
the grain nucleus and the other returning to the gas. Furthermore, nucleation
around positive metal ions is not favorable for these valence-bonded crystals, but
homogeneous nucleation is necessary for the first grain species to form and surface
nucleation can be utilized for later species. Nevertheless, the most important features
of classical nucleation theory are preserved; these can be summarized schematically
as follows.

Let r_s be the rate at which an appropriate atom (or molecule) from the gas hits
and sticks to a particular surface site on the grain (or droplet) nucleus; let t_c be the

cooling time [just after the temperature $T(t)$ falls below the condensation temperature T_{vap}] and let $\eta \gg 1$ be a dimensionless parameter defined by

$$t_c \equiv T_{vap} \left(\frac{dt}{dT}\right)_{T_{vap}} \quad , \qquad \eta \equiv r_s t_c \quad . \tag{1}$$

For homogeneous nucleation, only very small grain nuclei form with low abundance when the temperature T first falls below T_{vap}, but the growth of nuclei speeds up rapidly when a critical value ΔT of the difference $T_{vap} - T$ is reached. The object of nucleation theory is to calculate ΔT (a measure of the required supercooling), the number of particles N_{crit} in the critical nuclei at the instant when growth becomes important, and the number of particles N_{final} in a typical grain after most of the appropriate gas atoms have condensed out around the nuclei. The form of the expressions for ΔT, N_{crit}, and N_{final} depends on the form for the binding energy per particle. In classical nucleation theory, it is assumed that the binding energy $B_{N, N+1}$, released when the $(N+1)^{th}$ particle is added to a grain containing N particles, is of the form

$$B_{N, N+1} = B - N^{-1/3} D \quad , \tag{2}$$

where B and D are constants (which may depend on temperature but not on N).

When $B_{N, N+1}$ is of the form given in equation (2) and if $\eta \gg 1$, approximate analytic expressions for ΔT and for the order of magnitude of N_{crit} and N_{final} can be found. These classical nucleation theory formulas are

$$\frac{\Delta T}{T_{vap}} \approx \left[\frac{D}{B} \left(\frac{D}{2kT_{vap}}\right)^{1/2}\right] (\ell n\, \eta)^{-1/2} \quad ; \tag{3}$$

$$N_{crit} \sim (\ell n\, \eta)^{3/2} \quad ,$$

$$N_{final} \sim \ell n\, \eta \times \eta \quad . \tag{4}$$

The uncertainties for ΔT lie in the expression in square brackets, but its value should be of the order of, but somewhat smaller than, unity. The uncertainty in N_{final} lies in the coefficient multiplying η, but this coefficient, like $\ln \eta$ in equation (4), should be large compared with unity and small compared with η.

In carbon-rich stars (all O in CO, excess C in the gas), carbon-rich particles condense out well ahead of any other solids in a cooling sequence from 2000 to 1500 K. These particles should have density and latent heat similar to those of graphite, but they probably contain a little hydrogen, and the details of their optical properties and crystallographic structure are not yet known. Homogeneous nucleation is required, and the expression in square brackets in equation (3) is about 0.2. The gas-phase molecules are not monoatomic but C_2, C_3, or C_2H_2, so there are no heat-transfer difficulties (Salpeter, 1973). In oxygen-rich stars, the cooling sequence is more complicated, since minerals requiring less abundant constituents (Al_2O_3, alone or combined with calcium silicate) have a higher condensation temperature T_{vap} than does the more abundant magnesium silicate. Hence, the aluminum compounds probably undergo homogeneous nucleation first (with H_2O providing the required polyatomic molecule) and then provide surfaces for the magnesium silicates to undergo surface nucleation. Under typical giant-atmosphere conditions, the value of η for either graphite or the silicates would be $\sim 10^6$ or 10^7, but densities are lower in grain-producing atmospheres, and η is likely to be only of order $\sim 10^3$ or 10^4. Homogeneous nucleation will not work at all unless $\eta \gg 1$, so that very rare substances, such as tungsten, will not nucleate first even though they have larger values of T_{vap} than do the common minerals. For both carbon-rich and oxygen-rich stars, metallic-iron grains can condense out on the surface of existing graphite or silicate grains if the temperature becomes sufficiently low (see Figure 1).

Although we are interested primarily in growing grains, these grains may be streaming through the gas above the photosphere with rather high velocities, so we must also be concerned with destruction mechanisms. Sputtering by individual gas atoms or molecules is unimportant when the drift velocity V_{dr} of a grain, relative to the gas, is $V_{dr} \lesssim 10$ km sec^{-1} but is very destructive for $V_{dr} > 50$ km sec^{-1} (helium atoms and CO molecules being most efficient). Collisions between two grains with relative velocity V_{rel} are not well understood, but destruction (by shattering and/or partial evaporation) is probably unimportant when $V_{rel} \lesssim 1$ km sec^{-1} and fairly severe when $V_{rel} \gtrsim 5$ km sec^{-1}. The sticking probability for coalescence is particularly uncertain, but it is likely to be small compared with unity, especially for $V_{rel} \gtrsim 1$ km sec^{-1}. Small grains drifting slowly through gas that streams out of a stellar atmosphere may grow further by coalescing with other (slower or faster) small grains. There may even be some tendency to produce a uniform grain size, since grains that are too large may move so fast that they evaporate partially rather than coalesce further.

3. MASS FLOW FROM STELLAR ATMOSPHERES

According to equation (4), typical (graphite or silicate) grains would grow by condensation from the gas to a size of $N_{final} \sim 10^5$, corresponding to a radius a of order ~ 100 Å. The optical properties of these grains now become important in determining the force due to radiation pressure. Let z be the fraction of the total mass of the atmosphere in the form of grains and let Q be the average ratio of optical-extinction cross section to geometric cross section of the grain. The opacity coefficient κ in cm^2 g^{-1} of <u>total material</u> (grains plus gas) is related to Q, z, and the average grain radius a by

$$\kappa \propto \frac{zQ}{a} \quad . \tag{5}$$

For the simplest type of materials, Q is proportional to (a/λ) for absorption and to $(a/\lambda)^2$ for scattering. Figure 2 shows the actual, more complicated dependence of λQ on λ for interstellar dust grains. Nevertheless, it seems reasonable to assume that absorption dominates for the small grains and that Q is roughly proportional to a (with $Q \ll 1$ throughout), so that κ is roughly independent of a.

One parameter that characterizes a stellar atmosphere is the effective temperature T_e, which is close to the gas temperature near the photosphere. For grains to condense, the gas temperature has to be slightly less than T_{vap} (~ 1200 to 2000 K), and we consider only stars with $T_e < T_{vap}$, in which grains can form near the photosphere. On the other hand, such low temperatures are rare for stellar surfaces, so we can restrict ourselves to cases where T_e/T_{vap} is smaller than unity by only a modest factor. The total optical depth τ of the stellar atmosphere due to grains is of order $(T_{vap}/T_e)^4$, and we shall simply assume that $\tau \sim 1$. A second parameter that characterizes a stellar atmosphere is the surface gravity g or, more pertinent for our cases, the ratio of the upward acceleration due to radiation pressure to g acting downward. Assuming good coupling between the grains and gas, this ratio equals L/L_{crit}, where L is the actual luminosity of the star, and

$$L_{crit} \equiv \frac{4\pi c\,GM}{\kappa} \sim \left(\frac{M}{M_\odot}\right) 10^3 \, L_\odot \quad , \tag{6}$$

with M the stellar mass. If we were dealing with pure grains without any gas, the opacity would be κ/z, and L_{crit} would be replaced by zL_{crit}.

The dynamics of the atmosphere depends significantly on L/L_{crit}, and appreciable mass loss is possible only when this ratio exceeds unity. For a typical cool giant atmosphere, the thermal speed U_{th} and the escape velocity V_0 (in the absence of radiation pressure) are roughly

$$U_{th} \sim \left(\frac{kT_e}{\mu H}\right)^{1/2} \sim 3 \text{ km sec}^{-1} \quad ,$$

$$V_0 \sim \left(\frac{GM}{R}\right)^{1/2} \sim 30 \text{ km sec}^{-1} \quad . \tag{7}$$

The quantities of greatest interest are the velocity v_{gas} with which the gas emerges far from the star, the drift velocity V_{dr} of typical grains relative to the gas, and the total mass-flow rate ϕ (gas plus grains). The formulas are particularly simple if $L/L_{crit} \sim 2$ and if $Q \ll (U_{th}/V_0)^2 \sim 10^{-2}$, namely,

$$v_{gas} \sim V_0 \quad , \qquad V_{dr} \sim \left(\frac{QV_0}{U_{th}}\right) V_0 \quad ,$$

$$\phi \sim \frac{L_{crit}}{V_0 c} \quad . \tag{8}$$

The situation is more complicated for general values of $L/L_{crit} > 1$, but over a considerable range of this parameter,

$$\phi \sim \frac{(L\ L_{crit})^{1/2}}{V_0 c} \left(1 - \frac{L_{crit}}{L}\right)^{1/2} , \tag{9}$$

and both velocities v_{gas} and V_{dr} increase monotonically (but slowly) with increasing L.

Wickramsinghe (1972) has pointed out that grains would reach very high speeds if ejected from a stellar surface in the complete absence of any gas. In reality, as pointed out by Gilman (1973) and suggested by our equation (8), $V_{dr} \ll v_{gas}$ whenever there is an appreciable mass outflow, which occurs when $L > L_{crit}$. An interesting but complicated case arises when $zL_{crit} < L < L_{crit}$, so that grains could be ejected by radiation pressure if no gas were present but not if the grain/gas abundance ratio remained constant. In such cases, the gaseous atmosphere is "almost static," the grains drift upward with increasing drift velocity as the gas density decreases with increasing height, and eventually the grain/gas abundance ratio increases. There is some mass outflow of grains ϕ_{grain} in such cases, but it is much smaller than in equations (8) or (9), and the outflow rate ϕ of gas is much smaller still than $z^{-1}\phi_{grain}$. The flow rates, assuming again that $Q \ll (U_{th}/V_0)^2 \ll 1$, are roughly

$$\left(\frac{L_{crit}}{L}\right)^2 \phi \sim \frac{1}{z}\phi_{grain} \sim \frac{L}{U_{th} c} Q , \tag{10}$$

probably too small to be of practical importance. However, it is at least of academic interest to note that what few grains escape do so with <u>greater</u> velocity V_{dr} than that given in equation (8): Although the upward force on a grain is smaller for $L < L_{crit}$, the very much reduced gas-flow rates provide a much smaller drag on the grain.

Typical mass-flow rates ϕ in equation (8) for $L \sim 2L_{crit}$ are of order 10^{-6} to 10^{-5} $M_{\odot}$ year^{-1}. These rates are quite appreciable and are ~ 100 times larger than the rate at which hydrogen is converted into helium in the interior of the star. On the other hand, mass-flow rates ϕ in special kinds of stars, such as novae or stars ejecting planetary nebulae, can be even higher: An upper limit on ϕ is $\sim L/V_0^2$, which is larger by c/V_0 than the expression in equation (8). The quantity η in equation (1) is smaller in such cases of spontaneous outflow, and it is <u>not</u> understood theoretically how the nucleation of grains takes place; nevertheless, Woolf (1974, this volume) presents observational data to suggest that grains do indeed form when the outflowing gas stream has become sufficiently cool!

ACKNOWLEDGMENTS

I am grateful to F. Whipple for providing the stimulus for this symposium and to N. Woolf for making his manuscript available. This work was supported in part by National Science Foundation Grant GP36426X.

REFERENCES

Donn, B. D., and Stecher, T. P., 1965. Astrophys. Journ. <u>142</u>, 1681.

Donn, B., Wickramsinghe, N., Hudson, J., and Stecher, T., 1968. Astrophys. Journ. <u>153</u>, 451.

Fix, J. D., 1969. Mon. Not. Roy. Astron. Soc. <u>146</u>, 37.

Fix, J. D., 1969. Mon. Not. Roy. Astron. Soc. <u>146</u>, 51.

Gilman, R. C., 1973. Mon. Not. Roy. Astron. Soc. <u>161</u>, 3P.

Hoyle, F., and Wickramsinghe, N. C., 1962. Mon. Not. Roy. Astron. Soc. <u>124</u>, 417.

Salpeter, E. E., 1973. Journ. Chem. Phys. <u>58</u>, 4331.

Wickramsinghe, N.C., 1972. Mon. Not. Roy. Astron. Soc. <u>159</u>, 269.

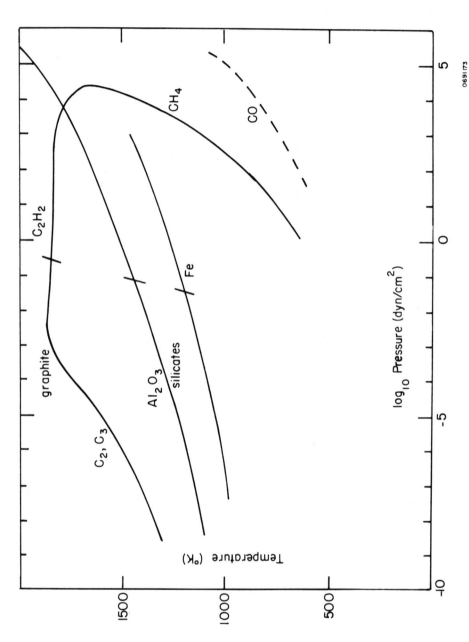

Figure 1.   A schematic phase diagram with temperature in °K and gas pressure in cgs units.   The slanting marks on each of the solid curves indicate where radiation and gas pressures are equal.   The labels C_2H_2, CH_4, and C_2, C_3 denote the most abundant carbon-carrying molecules in the relevant pressure range.

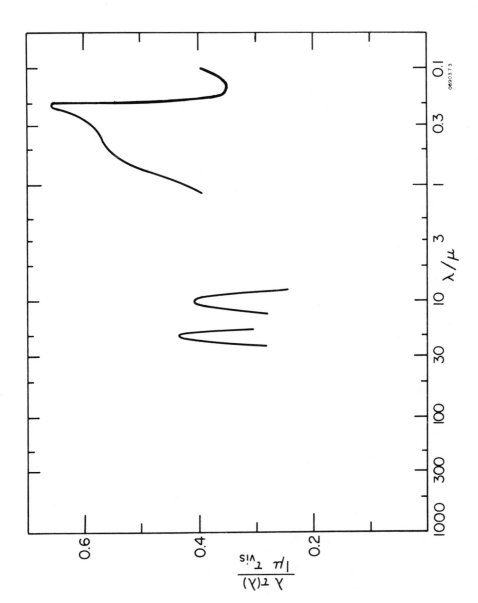

Figure 2.   A schematic plot of wavelength λ times optical depth τ in convenient units (proportional to λQ) versus λ for typical interstellar dust grains.

CIRCUMSTELLAR DUST

Neville J. Woolf

University of Minnesota, Minneapolis, Minnesota

ABSTRACT

The infrared spectral features of circumstellar solid matter are described. An attempt is made to identify the particles and to predict their sizes.

1. INTRODUCTION

In this paper, we shall be looking at evidence about the nature and formation of circumstellar dust particles. In previous papers (Woolf, 1973, 1974), I have considered the significance of this dust for its role in the death of stars and in the population of interstellar space, and I do not wish to consider these matters further at this time.

First we will briefly survey the circumstellar conditions under which dust is found. There is the circumstellar dust of the sun – the only low-luminosity, old, main-sequence star we know to have any dust. It has a zodiacal cloud, comets, and even planets with dust storms. Yet if similar dust existed around all stars like the sun, we would see no observable effects.

Instead, those stars we see with large amounts of circumstellar dust fit one of four categories:

1) Evolved cool giant stars with a large ratio of luminosity to their mass.
2) Evolved giant stars that oscillate violently.
3) Evolved hot stars that have shed a shell of now-ionized gas.
4) Young stars of all masses.

The dust may be known by its thermal emission, or by its time-variable blotting out of starlight, or by its modification of the spectral composition of starlight. If the dust is situated asymmetrically around the star, the starlight may be polarized and as the dust moves, the polarization may vary. Some dust clouds can be resolved against the sky, since they scatter starlight. If the central star is very hot, it is also possible that the dust may fluoresce. To the best of my knowledge, this possibility has never been checked by high-resolution spectroscopy. An obvious candidate is the visual continuum of the Orion Nebula.

All four groups of stars are emitting dust into space, but while we know that the old stars must be manufacturing dust, it is not clear that the young stars are not merely collecting and then reejecting the interstellar dust. Alternatively, if the material gets really hot, it is possible that the dust would be appreciably modified. The relative importance of such processes depends on the efficiency of converting interstellar matter into stars, and our lack of knowledge of this efficiency leaves it possible that such processes are either dominant or trivial in affecting the nature of dust in space.

The content of this review is affected by our ability to interpret infrared energy distributions. We have made some headway in interpreting the spectra of evolved stars but very little regarding young stars. This is partly because the dust shells around old stars tend to be optically thin and so can be interpreted with simple models, whereas the shells of young stars tend to be much thicker. Part of the problem is that we can more easily make models of matter leaving evolved stars. We know that the matter is gaseous, that it carries little angular momentum, and that its density far from the star will be negligible. None of these conditions is likely to be true for young stars. So, whereas most readers are probably interested in young stars that are forming planetary systems, I shall mainly address myself to dusty old stars. It is like the man who was looking for his door key under the streetlight: When asked where he lost it, he said he didn't know, but this was the only place he had a chance of finding it.

2. EVOLVED NORMAL-COMPOSITION COOL GIANT STARS

Evolved cool giant stars can be divided into two main groups. In the first, chemical abundances of the elements are rather similar to those in the sun, except occasionally for the ratio of hydrogen and helium to all heavier elements. In the second and far smaller group, the oxygen abundance is reduced below the carbon abundance. Other elemental abundances (except nitrogen, which plays no major role in dust) are relatively normal. The circumstellar dust grains in these two groups of stars have different infrared spectra.

The normal-composition stars have, in addition to the expected infrared continuum of the stars, emission peaks in the 10- to 20-µ spectral region. The emitting material has been identified as silicate dust (Woolf and Ney, 1969; Gilman, 1969). The spectra of Figure 1 are from Treffers and Cohen (1973). The stars show two emission peaks near 1030 and 550 cm^{-1}. In α Ori, the material seems to be optically thin. In VY CMa, the 1030 cm^{-1} peak seems flattened as though it is becoming optically thick, and for W Hya, the peak is even flatter. Possibly there is an additional emission peak in this star near 770 cm^{-1}.

The material seen here seems to be cosmically common. Figure 2 shows the emission spectrum of the center of the Orion Nebula, which for this purpose should be considered as the circumstellar cloud of a group of hot stars. The emission has approximately the same spectral shape as Figure 1. Figure 3 is an absorption spectrum of an object of unknown nature near the center of our Galaxy. Here again, the opacity wavelength variation is rather similar. However, there seems to be an opacity minimum near 780 cm^{-1}, whereas for Figure 1 this minimum is at about 750 cm^{-1}, and in Figure 2 it seems to be shortward of 740 cm^{-1}. Such slight differences could be produced by different size distributions for the emitting or absorbing dust.

A similar spectral feature was seen, in emission, in Comet Bennett 1969i (Maas, Ney, and Woolf, 1970; Ney, 1970; Hackwell, 1971). The spectrum, Figure 4, was obtained under difficult conditions. Nonetheless, it seems that this material is quite similar to that of the objects outside the solar system. Observations of Comet Kohoutek 1973f made since the symposium confirm this.

Figure 5 compares the emission peak of one cool star and an inverted absorption spectrum of a Type II carbonaceous chondrite meteorite. Another comparison is the spectrum of dust on Mars (Hanel et al., 1973) shown in Figure 6. The Martian dust is quite distinct from this other "standard" dust. Table 1 shows the key features of the spectra.

Figure 7 shows these frequencies for various siliceous rocks studied by Lyon (1964). Neither Martian nor standard dust exactly fits the laboratory materials. However, the Martian dust seems relatively acidic, whereas the standard dust seems relatively basic. Such behavior can be understood as that expected from materials that have and have not undergone water erosion, respectively.

In both cases, it seems that the high-frequency band is stronger than the low-frequency one. Such behavior is common but not invariable among silicate rocks.

The identification of standard dust as silicates is, of course, less certain than most astrophysical identifications from many atomic or molecular lines. In part, from circumstantial evidence, it appears that

1) The matter must incorporate a large fraction of the heavy elements to be so abundant.

2) The matter is highly nonvolatile, being found near the surface of stars, and it survives ultraviolet-radiation bombardment and a low-density environment near the Orion Nebula O stars.

3) It occurs in a comet; comets in turn produce shower meteors. Shower-meteor spectra seem to indicate the presence of heavy elements in roughly those proportions present in silicates.

A number of apparently normal-composition stars have an approximately blackbody spectrum, apart from the emission peaks, that corresponds to temperatures below 1500 K. Gas at these temperatures has very little continuous opacity, so it is generally assumed that the opacity is due to dust. However, dielectric solids are good scatterers, but poor at absorbing a continuum, so it must be assumed either that we have failed to discover some type of atomic or molecular continuum or that there is a metallic or semiconductor component to the dust. Iron is one metal found in solid form in the solar system and this could be the material, but it has no distinguishing spectral features. It is also possible that the dust is so damaged by the continual bombardment of gas atoms and molecules that the continuous opacity is due to broken silicate bonds. The bond-breaking and annealing rates could perhaps be calculated to distinguish between these possibilities.

3. CARBON-RICH GIANT STARS

Carbon-rich, dust-ejecting stars are probably 10 times as rare as normal dusty stars. The spectra of these stars show a single emission peak near 930 cm^{-1} discovered by Hackwell (1972). Tentative initial identifications were Si_3N_4 and SiC. Gilman (1969) demonstrated that at the pressures typical of stellar atmospheres, the first condensates that occur as carbon-rich matter cools are carbon itself and then SiC. Hackwell has shown that this is still true in circumstellar space at pressures ~5×10^{-5} dyne cm^{-2} (see Figure 8).

Treffers and Cohen (1973) have shown that the carbon star +10 216 has the energy of its 930-cm^{-1} peak almost entirely confined to the region 790 to 980 cm^{-1}, as seen in Figure 9. These frequencies are the two defining phonon frequencies for SiC, which seems to provide convincing proof of the identification. Figure 10 (from Lyon, 1964) shows a laboratory spectrum for comparison.

The smooth continuum of this star closely resembles a blackbody at a temperature of about 600 K. Again this seems to be evidence for metallic or semiconducting matter. In this case, various forms of carbon could possibly condense, and the continuum may be the evidence that this has happened.

The violently pulsating but rare RV Tau stars also have some carbon-rich members; two of these, AC Her and RU Cen, have been photometrically observed by Gehrz and Ney (1972), but they are rather faint and have not yet been observed spectroscopically. Figure 11, taken from their paper, shows energy distributions for these stars unlike those known for other objects; they include two long-wavelength peaks, so that the spectra resemble those of normal stars rather than of cool carbon stars. However, the short-wavelength peak seems shifted to about 1100 cm^{-1}, and the long-wavelength peak is unusually prominent. The detailed chemical composition of these stars is not known. Fortunately, objects like these are extremely rare, so the production of this exotic dust does not play a major role.

A group of hot carbon stars are also very deficient in hydrogen. Some of these stars have erratic declines in their light output and are called R CrB variables. They also show a strong blackbody-like infrared continuum (Feast and Glass, 1973). The other stars in the group do not vary appreciably, nor do they have a strong infrared continuum. The standard interpretation is that all these stars are single stars that have lost their outer layers. For some unknown reason, the R CrB stars, but not the others, occasionally condense carbon dust in their outer layers.

An alternative hypothesis now being discussed is that this group is composed of very long-period binary stars that have undergone mass exchange, and that the R CrB variables are those stars in which the companion is a cool giant star. Clumps of dust in the extended outer layers of the cool star occasionally occult the hot star.

I will not now discuss the different predictions of these two hypotheses. But if the first is correct, then as we watch the light of the hot star decline, we should learn about the condensation of dust; whereas, with the second hypothesis, we learn about the clumping of matter in circumstellar shells.

The spectral shape of the infrared continuum of R CrB is of a somewhat flattened blackbody with a characteristic temperature varying between 900 and 500 K with, perhaps, a periodicity of 1250 days. No spectral peaks are discernible in the 10- to 20-μ region. Some of the R CrB stars are B stars. If these are condensing dust, it must condense very far from the surface of the star.

3.1 Planetary Nebulae

Many planetary nebulae show an infrared continuum that seems to thermalize a substantial fraction of the energy emitted by the ultraviolet-hot central star. To a first approximation, the continuum seems like a blackbody; however, Gillett, Forrest, and Merrill (1973) have shown that in two of these nebulae, the continuum appears to be increasing toward the short-wavelength end of the 8- to 13-μ region and to have a moderately sharp spike near 11.3 μ (Figure 12). One possibility is that this material is composed of carbonates (see Figure 13). No theoretical basis exists for this identification, but such sharp features produced by solids are rare, and the carbonate radical does indeed fit the spectral behavior. Conclusive evidence should be found when planetary nebulae can be observed in the 5- to 8-μ range. Possibly, there is a different sequence of condensates when ionized material condenses.

3.2 Young Stars

Not all young stars have spectra that are difficult to interpret. A number of them are surrounded by a thin layer of standard dust at typical temperatures ~200 K. But many of the more interesting objects are apparently in clouds so dense that the spectral features have vanished. Figure 14 shows energy distributions for a number of these objects.

T Tau is one of the few stars of this kind that has been observed thoroughly in the infrared. M. Cohen (1973, private communication) reports evanescent silicate features seen in the 10- and 20-μ spectral bands. For a number of these stars, it seems that the absorbing matter forms a dense equatorial belt that confines visual emission to two polar cones. As a result, the more distant dust is illuminated to form conical or biconical nebulae with reflection spectra.

These stars present a variety of problems to the interpreter, one of the worst being their inconsistency. For example, spectroscopic reports sometimes indicate that matter is being ejected and, at other times, that it is falling in. One gets the impression that infall is a relatively rare phenomenon, perhaps confined to a small fraction of the stars.

4. CIRCUMSTELLAR CONDENSATION

At least for some objects, dust condenses in low-density regions well outside the photospheres of the stars. A first indication that this is so comes from the planetary nebulae. Central stars have surface temperatures of tens or hundreds of thousands of degrees Kelvin. The dust is unlike that ejected from possible earlier phases as cool stars and so must condense from the circumstellar gas during the planetary nebula phase. Also, when dust is seen condensed around stars as hot as type G0 supergiants, ~5500 K, it is hard to imagine that there are star spots on the surface with temperatures six times cooler than average; even if such spots do occur, it is unlikely that dust could escape vaporization when it leaves the spot and is exposed to the average radiation field at some distance above the surface.

R. Gilman (1973, private communication) has estimated the distance above a stellar surface at which particles of various compositions could persist in a gas of pressure 5×10^{-5} dyne cm^{-2}. The only heating mechanism assumed is the star's radiation. Figure 15 shows that the above-mentioned problem of silicates around a G0 star is severe. Figures 15 and 16 show that some deviations from the sequence predicted in Figure 7 do occur. Thus, for example, in the region around a cool carbon star, SiC will condense before carbon itself does, even though it evaporates at a lower temperature; for normal-composition stars, silicates can condense far closer to the surface than can iron. These calculations confirm the impression that, for many stars

that appear to have dust envelopes, matter can only condense when departing gas
has reached a few stellar radii above the surface.

I shall ignore the problem of nucleation, merely noting in passing that nature seems
to achieve with ease what seems nearly impossible to us. And so we turn to the problem
of particle growth. Radiation pressure forces the dust grains through the gas at ter-
minal velocity with two results: 1) The momentum from the radiation carries the gas
away from the star in an expanding envelope, and 2) the dust is able to grow by its
collisions with the gas atoms. As the dust grows, however, the radiation pressure
forces the dust to higher and higher velocities. Eventually the gas atoms start sputtering the dust, and an equilibrium radius for the dust particles is reached when as much
matter leaves the particle per second as tries to stick. Alternatively, if the particle
velocity never becomes sufficiently high, the dust will cease to grow because the
particles run out of gas.

Gilman has also calculated radiation-pressure effective cross sections for the
different particles, and with these results, I have estimated the final particle radii.
In Figure 17, the final particle size attained is shown as a function of mass-loss rate.
The particles shown are graphite, assumed to form in a carbon-rich star, and
silicates, assumed to form in a normal-composition star. The star is assumed to be
a 2500 K giant, of 5×10^3 solar luminosities. The typical stars contributing to the
interstellar medium are similar to this and are losing about 10^{-6} $M_\odot$ yr^{-1}. At this
rate, the silicate particles formed are about 10^{-5} cm in radius, and the carbon
particles are about 2×10^{-6} cm. The SiC particles should be of similar size to the
silicates, whereas iron should be similar in size to the graphite. High-luminosity
stars will typically produce particles a few times smaller than this, but they contribute a smaller fraction of matter to the interstellar medium.

The dust grows, in the absence of sputtering, so that half the mass of a grain is
added beyond five times the condensation stellar radius. In such cases, the matter
is accumulated at an average, relatively cool temperature of about 450 K. This would
apply to the silicate particles. But if sputtering is important, as it should be for
graphite, then the particles grow to their maximum size while they are almost at
their vaporization temperature of about 1700 K.

5. SUMMARY AND CONCLUSION

Observationally, there seem to be three main types of dust in circumstellar envelopes. Normal stars condense silicates and possibly iron. Carbon stars condense SiC and probably carbon. And planetary nebulae condense an unknown material that may be carbonates. The particle sizes are limited, both by sputtering and by running out of gas, with typical particle sizes expected to be about 10^{-5} cm in radius for dielectric materials and 2×10^{-6} cm for metallic and semiconductor particles. In many cases, it seems that the dust must condense in circumstellar space. It is possible that interstellar dust is substantially changed in the circumstellar envelopes of young stars, but this question depends on the unknown efficiency of star formation. The entire discussion of stars, other than the sun, has ignored the possibility of objects like comets or meteorites. This is because the phenomena seem confined in the H-R diagram in such a way that direct formation of dust has always seemed possible; the large amounts of gas seen leaving most of the stars make this more plausible, and it is unnecessary to assume that we are witnessing the breakup of comets or meteorites.

ACKNOWLEDGMENTS

This work was supported by the National Science Foundation under grant GP 32772. I am indebted to Drs. Martin Cohen and Robert Gilman for information on work in progress.

REFERENCES

Feast, M. W., and Glass, I., 1973. Mon. Not. Roy. Astron. Soc. 161, 293.

Gehrz, R. D., and Ney, E. P., 1972. Publ. Astron. Soc. Pacific 84, 768.

Gillett, F. C., Forrest, W. J., and Merrill, K. M., 1973. Astrophys. Journ. 183, 87.

Gilman, R. C., 1969. Astrophys. Journ. (Lett.) 155, L185.

Hackwell, J. A., 1971. Observatory 91, 37.

Hackwell, J. A., 1972. Astron. Astrophys. 21, 239.

Hanel, R., Conrath, B., Hovis, W., Kunde, V., Lowman, P., Maguire, W., Pearl, J., Pirraglia, J., Prabhakara, C., Schlachman, B., Levin, G., Straat, P., and Burke, T., 1972. Icarus 17, 423.

Lyon, R. J. P., 1964. NASA Contract Report CR-100.

Maas, R. W., Ney, E. P., and Woolf, N. J., 1970. Astrophys. Journ. (Lett.) 160, L101.

Ney, E. P., 1970. Sky and Tel. 40, 141.

Treffers, R., and Cohen, M., 1973, preprint.

Woolf, N. J., 1973. In Interstellar Dust and Related Topics, Proc. IAU Symp. No. 52, ed. by J. M. Greenberg and H. C. van de Hulst (D. Reidel Publ. Co., Dordrecht-Holland) p. 485.

Woolf, N. J., 1974. I.A.U. Symposium No. 66, in preparation.

Woolf, N. J., and Ney, E. P., 1969. Astrophys. Journ. (Lett.) 155, L181.

TABLE 1.

Comparison of Martian and standard dust.

	Low-frequency peak (cm^{-1})	Minimum (cm^{-1})	High-frequency peak (cm^{-1})	High-frequency cutoff (cm^{-1})
Martian dust	480	~800	1090	~1250
Standard dust	550	~750	1030	~1150

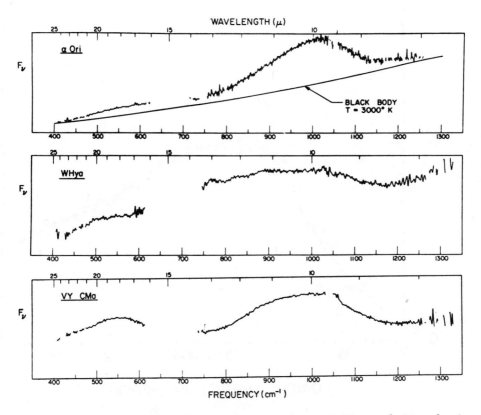

Figure. 1.   Spectral-energy distribution for normal-composition cool stars showing circumstellar dust (from Treffers and Cohen, 1973).

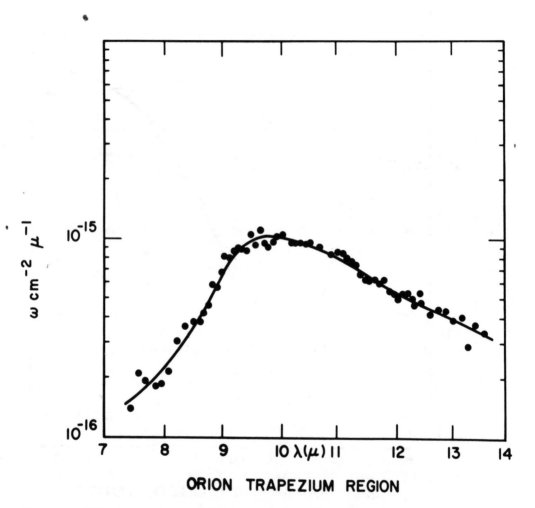

Figure 2.   Emission spectrum of the Orion Nebula near the Trapezium stars (from
            Gillett).

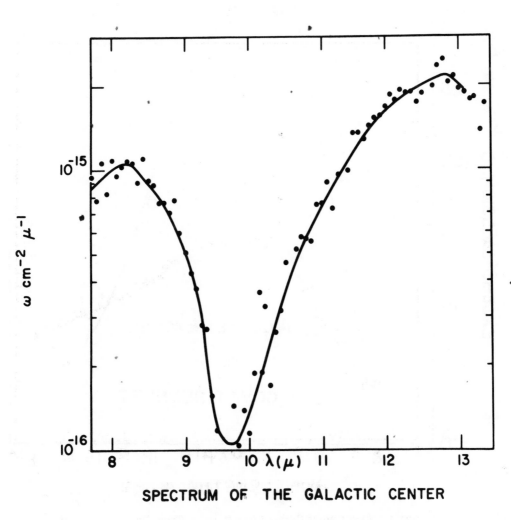

SPECTRUM OF THE GALACTIC CENTER

Figure 3.   Spectrum of midinfrared source near the galactic center.

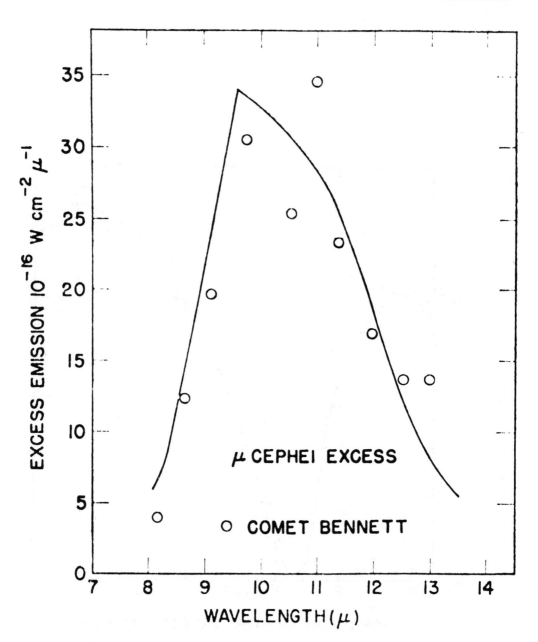

Figure 4. Spectrum of Comet Bennett 1969i (from Hackwell).

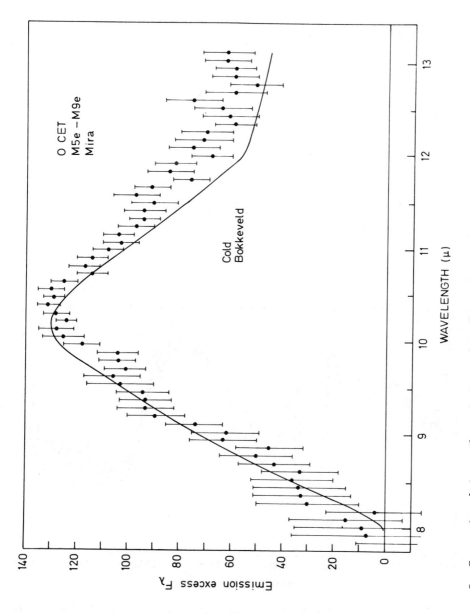

Figure 5.   Comparison between the emission peak in the spectrum of o Cet and the absorption spectrum of a carbonaceous chondrite (from Hackwell).

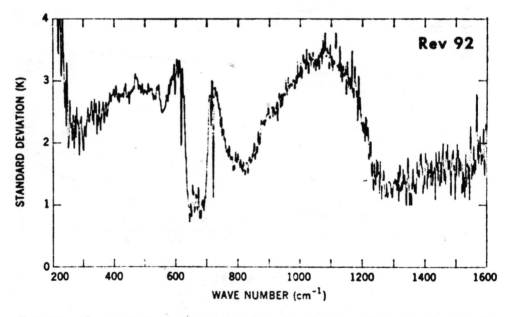

Figure 6.   The spectrum of Martian dust (from Hanel et al., 1972).   The features 550
to 800 cm^{-1} are caused by Martian CO_2.

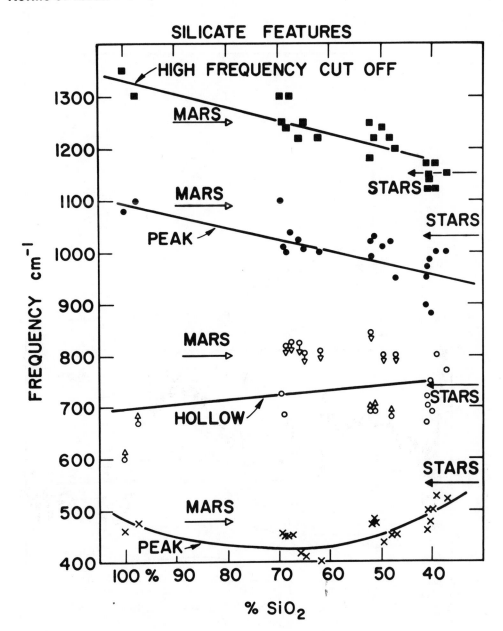

Figure 7. Key spectral features for silicate dust as a function of rock acidity.

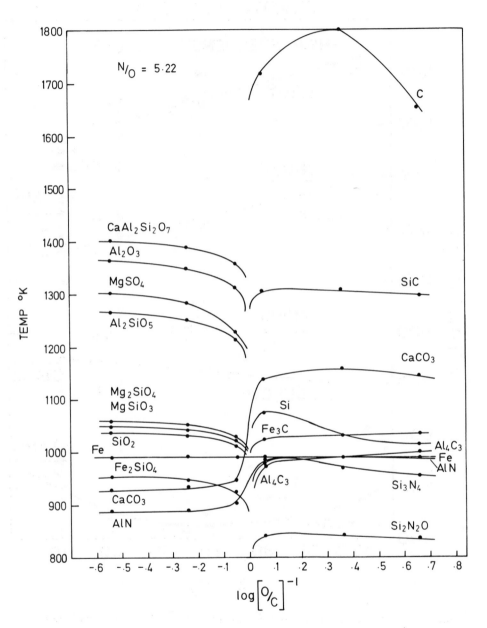

Figure 8. Condensation sequences for circumstellar gas of various O/C ratios (from Hackwell).

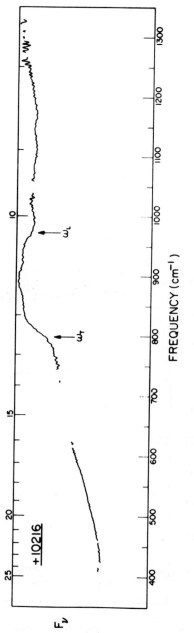

Figure 9.   Spectral-energy distribution for the carbon star +10 216 (from Treffers and Cohen, 1973).

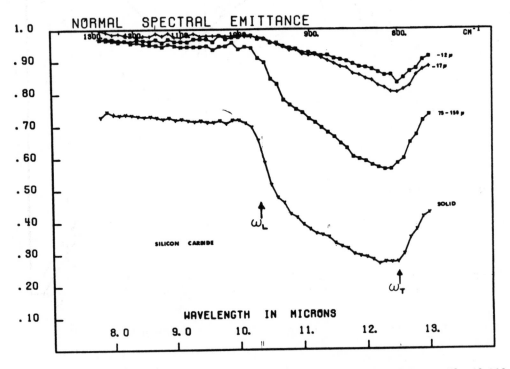

Figure 10. Laboratory spectrum of SiC (from Lyon, 1964) for comparison with +10 216.

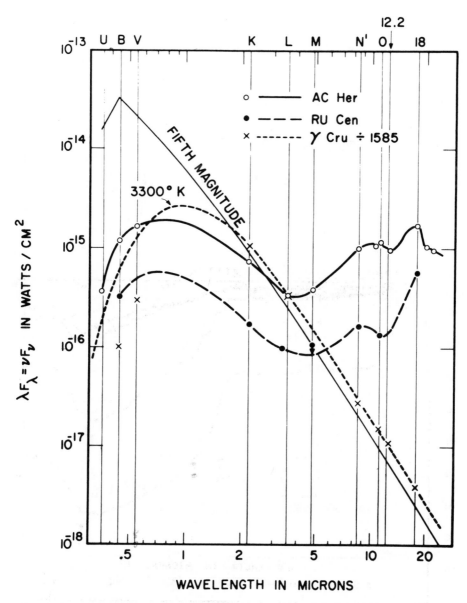

Figure 11.   Energy distribution for two carbon-rich RV Tau pulsating variable stars
(from Gehrz and Ney, 1972).

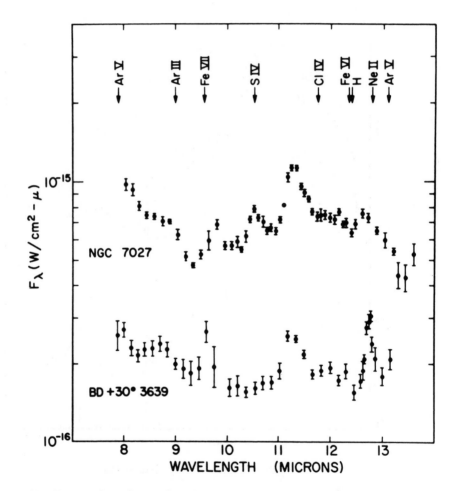

Figure 12.   Energy distribution for planetary nebulae (from Gillett, Forrest, and
              Merrill, 1973).

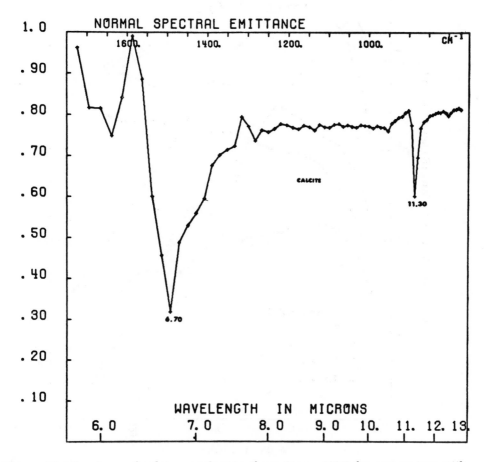

Figure 13. Spectrum of calcium carbonate (from Lyon, 1964) for comparison with NGC 7027.

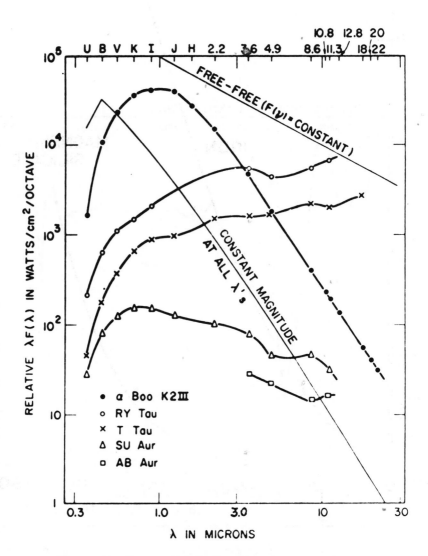

Figure 14. Energy distribution for some young stars.

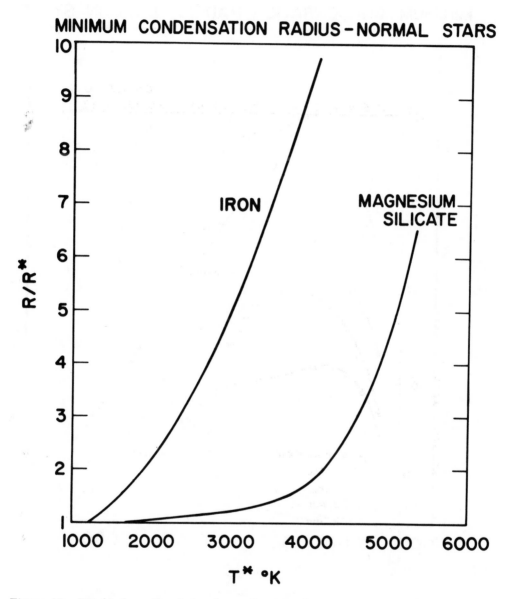

Figure 15.   Height above the photosphere of a normal-composition star for dust to
condense (adapted from Gilman, 1969).

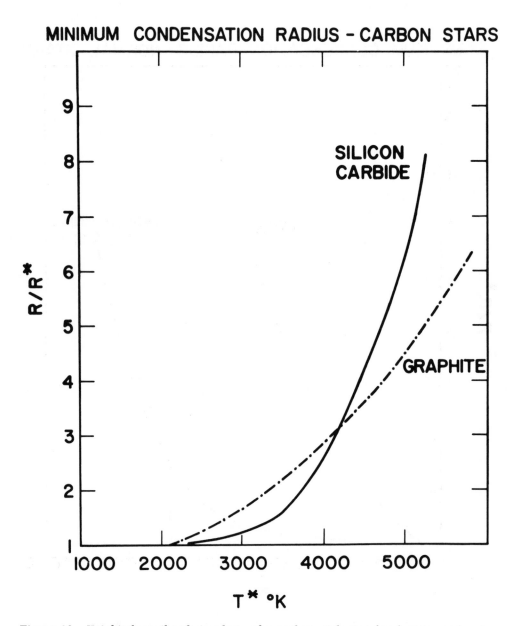

Figure 16.   Height above the photosphere of a carbon-rich star for dust to condense
             (adapted from Gilman, 1969).

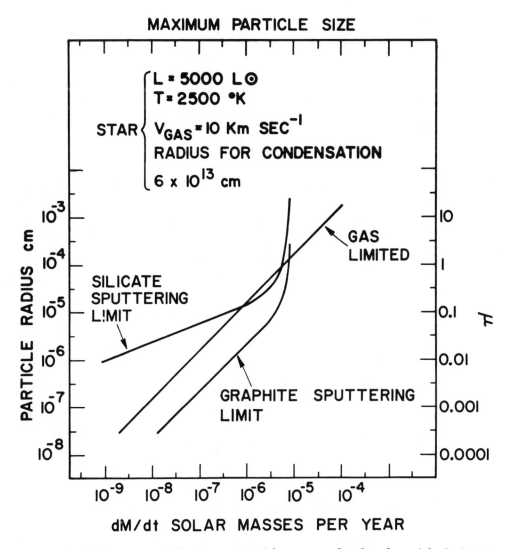

Figure 17.   Maximum particle sizes emitted from normal and carbon-rich giant stars.

THE COMPOSITION OF INTERSTELLAR DUST

George B. Field

Center for Astrophysics
Harvard College Observatory and Smithsonian Astrophysical Observatory
Cambridge, Massachusetts

ABSTRACT

Direct evidence that interstellar dust is composed partly of silicates, graphite, and water ice is reviewed. Indirect evidence, from recent studies of the chemical composition of interstellar gas, is assessed in terms of two possible models for the formation of the dust: condensation under thermal-equilibrium conditions and accretion under nonequilibrium conditions. It is concluded that probably the more refractory elements condense under equilibrium conditions and that probably the more volatile ones condense under nonequilibrium conditions. Equilibrium condensation may occur either in stellar atmospheres or in circumstellar nebulae, but arguments from stellar evolution favor the latter. If this is correct, all but a tiny fraction of the present interstellar medium has at least once been involved in circumstellar nebulae. This is consistent with the hypothesis that planetary systems are commonplace.

1. DIRECT EVIDENCE OF THE COMPOSITION OF INTERSTELLAR DUST

Since this symposium, like the scientist it honors, embraces a number of different fields, it may be helpful to review briefly our present knowledge about interstellar dust.

To the stellar astronomer, interstellar dust is a nuisance because it attenuates the light from stars in the galactic disk. To the specialist in the interstellar medium, however, it presents an intriguing unsolved problem, in that the composition and origin of the dust are still not completely understood after half a century of investigation. Here I will argue that interstellar dust in the Galaxy (which is comprised of about 100 million solar masses) started its life in clouds of gas surrounding newborn stars (Dorschner, 1968; Herbig, 1970). Studies of interstellar dust may therefore provide information on events similar to those that led to the formation of the solar system.

The main source of information about interstellar dust is the extinction curve, which describes the attenuation of starlight as a function of wavelength. Until recently, the extinction curve was known only between 0.3- and 1.0-μ wavelength, but recent extensions to the ultraviolet (0.1 μ) and far-infrared (20 μ) have given vital new data, including all the direct evidence on composition (Figure 1). The latter depends on characteristic resonances in the extinction curve, as we shall see below, but something can be said just from the total amount of extinction, which is observed to increase (irregularly) with distance, indicating a general distribution of dust. Purcell (1969), by applying the Kramers-Kronig dispersion relation to the extinction curve, has shown that at least 9×10^{-27} of the volume of space must be occupied by dust grains, if they are roughly spherical. Since hydrogen and helium cannot condense at interstellar temperatures, the dust must be composed of heavier elements, such as the abundant elements C, N, O, Mg, Si, S, and Fe. The solids that can be made from these elements (with hydrogen) have mass densities s in the range 1 to 10 g cm^{-3}, so that one can infer a mean smoothed-out mass density for the dust, which, when compared with the mass density of hydrogen known from 21-cm and ultraviolet studies, leads to the conclusion that the dust–gas ratio must exceed 0.5 to 5% by mass. Since, in the sun and nearby stars, all the heavy elements together comprise only 1.4% by mass, a substantial fraction of the heavy elements in interstellar space must be bound up in the dust. The problem then becomes: What chemical compounds and physical structures are involved, and what is then implied for the origin of the dust?

Van de Hulst (1949) argued that the chemical compounds in the dust would probably be hydrogen-rich, in view of the great cosmic abundance of hydrogen. He therefore proposed that the dust is composed mainly of the ices of the saturated compounds of C, N, and O (CH_4, NH_3, and H_2O), and he showed that a judicious choice of particle size (≈ 1000 Å) sufficed to explain the extinction curve quantitatively if significant fractions of the C, N, and O are in the dust. Small admixtures of the other abundant elements in unspecified compounds led to the concept of "dirty ice," rather similar to the material proposed by Whipple (1950) for the nuclei of comets. Indeed, a possible relationship between long-period comets (whose aphelia are at interstellar distances) and interstellar dust has long been the subject of speculation.

Van de Hulst postulated that atoms of heavy elements in the interstellar gas hit and stick to the dust grains, causing them to grow by accretion. At typical interstellar densities (1 to 100 H cm^{-3}), the grains grow to the required size in about 10^6 to 10^8 years, only 10^{-4} to 10^{-2} of the age of the Galaxy. The fact that a substantial fraction of the heavy elements are in the dust might be explained as a consequence of the exhaustion of gas-phase heavy elements (but not of hydrogen, which can condense only by forming compounds with much less abundant elements).

2. EQUILIBRIUM CONDENSATION

Van de Hulst's theory does not explain the presence of particles on which accretion can occur, and numerous attempts to explain their spontaneous formation under interstellar conditions have failed. O'Keefe (1939) suggested that such particles may form in the atmospheres of cool stars, where the density is enormously higher than in space, and he particularly advocated the formation of graphite flakes in the (rare) stars in which carbon is more abundant than oxygen (carbon stars). Dorschner (1968) and Herbig (1970) suggested that such a condensation process may occur, not only in stellar atmospheres but also in extended circumstellar nebulae like the solar nebula, which are inferred to be present around many young stars such as the T Tauri stars.

In both cases, radiation pressure on the dust has been invoked to drive the dust away from the star and into space.

Spitzer (1948) was among the first to demonstrate, at an earlier symposium at Harvard College Observatory, the major dynamical consequences of radiation pressure on dust grains, owing to the fact that the ratio of the outward radiation force to the inward gravitational force on a dust grain is largest for scattering particles with sizes close to the wavelengths of radiation emitted by the star. We shall hear more about the condensation in and escape of dust from stars from Dr. Salpeter (1974, this volume).

The condensation of dust grains in stellar atmospheres or in circumstellar nebulae is a complex matter. Nevertheless, equilibrium calculations, carried out by a number of authors for the case of the solar nebula (see papers by Drs. Lewis and Grossman, 1974, this volume), have succeeded surprisingly well in accounting for the observed solid matter in the solar system. Such calculations may be employed, therefore, to get a rough picture of the formation of interstellar dust grains in a stellar atmosphere or nebula. Later (Section 5) we shall argue that the actual condensation is primarily in nebulae, not atmospheres. Henceforth, we shall use the word "nebula" to refer to both types of region.

Consider such an atmosphere or nebula in which the temperature T is falling. In the case of an atmosphere, this might happen because the gas is rapidly moving away from the star under the influence of radiation pressure; in a nebula, the heat gained in gravitational contraction from an interstellar cloud could be radiated away.

At high temperature, all the elements z are monatomic gases, but as the temperature falls, they react to form a variety of molecular gases. For example, Grossman (1972) finds that if the total pressure is constant and equal to 10^3 dynes cm^{-2}, Si is divided between SiO (70%) and SiS (30%) below 1700 K. When the temperature falls to 1625 K, these gases react with Ca and Mg vapor, and with the solid corundum (Al_2O_3) (which has already condensed) to form melilite, a mineral containing Ca, Al, Si, Mg, and O. However, because Ca and Al are less abundant than Si, only a small fraction of Si is condensed this way. At 1440 K, however, SiO and SiS begin to react with Mg and O in the vapor phase to form forsterite (Mg_2SiO_4). Since Mg has the same abundance as Si and O is much more abundant than either, this uses up half the remaining Si and much of the Mg by about 1360 K. At 1350 K, enstatite ($MgSiO_3$) becomes stable, and it uses up essentially all the Si when T has reached 1200 K.

The point where each solid begins to form is called the condensation temperature of that solid. For each element, there is a condensation temperature of particular interest, namely, that where the element begins to be depleted quantitatively by condensation. In the case of Mg, this is 1440 K, corresponding to the formation of forsterite; for Si, it is 1350 K (enstatite). As we shall see, there is evidence that this process has occurred, judging by the gas-phase abundances in interstellar space. In determining those abundances, only fairly large changes (factors of 3 or more) are reliable, so the elemental condensation temperature is of greatest interest here.

Condensation temperatures have been calculated by many authors, most recently by Grossman (1972), Cameron, Colgate, and Grossman (1973), and Lewis (1972). Table 1 lists elemental condensation temperatures for the more abundant elements, appropriate for a total pressure of 10^3 dynes cm^{-2} (10^{-3} atm) in an oxygen-rich gas (O/C = 1.75).

Although oxygen first condenses as water, it later reacts quantitatively with CH_4 to form a hydrate. Similarly, although iron first condenses as the metal, it reacts first at 700 K with sulfur to give troilite (FeS), then with enstatite at 500 K to form fayalite (Fe_2SiO_4), and finally with water at 400 K to form magnetite (Fe_3O_4). Hence, one expects the most abundant elements to be in the form given in Table 2, if condensation occurs in oxygen-rich nebulae — namely, hydrates of ammonia and methane (with solid neon), possibly hydrocarbons, magnesium and iron silicates, metallic nickel-iron, troilite, and magnetite. We now consider the astronomical evidence concerning such condensates.

Figure 1 shows a conspicuous bump in the extinction curve at 2200 Å. This was predicted by Wickramasinghe and Guillaume (1965) and Stecher and Donn (1965) as a signature of graphite, and Gilra (1971) has shown that the data can be fitted very well with graphite particles about 250 Å in radius. Graphite does not condense in an oxygen-rich gas, but Gilman (1969) has shown that it is the first material to condense as the temperature is lowered in a carbon-rich gas (C/O = 2.0; p = 100 dynes cm^{-2}). Interstellar graphite could therefore be produced by condensation in carbon-rich objects, as discussed by Hoyle and Wickramasinghe (1962).

Data presented by Dr. Woolf (1974, this volume) show that both water ice and silicates have been detected in the infrared interstellar-extinction spectrum at 3.1 μ and 9.7 μ, respectively. As yet there is no spectroscopic evidence for iron, but it would help to explain interstellar polarization (Aannestad and Purcell, 1973), an effect that is produced by the alignment of interstellar dust grains in a magnetic field. Magnetite, with its ferromagnetic properties, may be a good candidate (Purcell, 1973). Thus, of the eight most abundant elements, there is at least some evidence that five are present in one of the forms predicted by equilibrium-condensation theory.

3. EVIDENCE FROM GAS-PHASE ABUNDANCES

Since the heavy elements that make up the dust grains are drawn from the gas phase, they should be depleted there. If the dust grains form in an equilibrium process, the most refractory elements should be most depleted, and the most volatile ones should be least depleted. This statement can be made quantitative. If the condensation of element z takes place in a medium of total molar density s and temperature T, via the formation of a condensed molecule, and if $\xi \equiv s(z)/A(z)s$ is the depletion factor by which the molar density $A(z)s$ expected on the basis of cosmic abundance $A(z)$ is reduced, then one can show (Appendix A) that

$$\log \xi = \log \left[\frac{T_c(m)}{T} \right] - \frac{\Delta G(m)}{2.3R} \left[\frac{1}{T} - \frac{1}{T_c(m)} \right] , \tag{1}$$

where $T_c(m)$ is the condensation temperature of the appropriate molecule (Table 1) and $\Delta G(m)$ is the corresponding change in free energy. This formula applies if $T < T_c(m)$; otherwise, $\xi = 1$. The condensation process is terminated at the temperature at which the gas and dust ejected from the vicinity of the star reaches such a low density that all kinetic processes effectively cease. One would then be left with a mixture of gas and dust "frozen" at that temperature. If this temperature is T, then elements with $T_c(m) < T$ would have $\xi = 1$, while elements with $T_c(m) > T$ would have $\xi < 1$, as given by equation (1). Since $\Delta G(m)/2.3R$ is typically ~20,000 K for the refractory elements, far larger than the values of $T_c(m)$ listed in Table 1, one expects severe depletion (several orders of magnitude) of those elements for which $T_c(m)$ considerably exceeds T. In actual interstellar clouds, where the dust has originated from different stars, this expected result will be smeared out because there are different values of T in different stars, but since the predicted effect is orders of magnitude, one might hope to detect it in the actual gas-phase interstellar abundances.

Alternatively, if interstellar dust forms primarily in interstellar space, via the nonequilibrium accretion process of van de Hulst, one would expect to find different dependences of ξ on z. Thus, if dust grains are negatively charged, positive ions such as C^+ and Na^+, (which are the dominant forms of C and Na in interstellar space) would be preferentially attracted to the dust grains. Mészáros (1972) has shown that in this case, $-\log \xi$ increases with time, the rate being proportional to $(1 + 2.5Z)A^{-1/2}$, where Z is the charge and A is the mass of the ion in question. On the other hand, if the dust is positively charged, ions would be totally repelled, and neutrals such as N and O would accrete at rates proportional to $A^{-1/2}$ (Aannestad and Purcell, 1973). Without further specification of the grain composition, little can be said, for it is likely that graphite and water ice would be negatively charged, while silicates would be positively charged (Watson, 1972; Feuerbacher, Willis, and Fitton, 1973).

Field (1974) has collected the data on gas-phase interstellar abundances of 19 elements in the cloud in front of the 09.5 V star ζ Oph, using ground-based, rocket, and OAO-3 data (7, 1, and 11 elements, respectively). Corrections were made for unseen ionization states (such as Ca^{++} and Na^+), by using an electron density of 0.07 cm^{-3} deduced from the Ca^+/Ca^0 ratio of White (1973), together with standard radiation fields and photoionization cross sections. Most (5 out of 7) of the elements observed from the ground required ionization corrections, which are the main source of uncertainty in their abundances, since the ground-based observations are of such high resolution that corrections for saturation are readily obtained. On the other hand, the observations made by the Copernicus satellite (OAO-3) are usually of the dominant ionization state, but are uncertain because low resolution does not permit accurate corrections for saturation. The uncertainties due to this cause were estimated by the authors (Spitzer et al., 1973; Morton et al., 1973) and are included in the data below.

The resulting abundances show interesting effects, among which are the following:

1) Relative to H, the abundances of some elements (S, Li) are nearly equal to those in the solar system (for which I employed the values of Cameron, 1973), while none is greater. This lends support to the methods of data analysis and interpretation.

2) Twelve elements are depleted; the rest (four elements) are uncertain enough that one cannot say. This is qualitatively in agreement with the expectation that most heavy elements are bound up in grains.

3) While some of the depleted elements may be in molecules, the evidence indicates that this effect is negligible. H_2 contains 2/3 of the hydrogen nuclei toward ζ Oph. Outside of that, only CO, CH, CH^+, and CN have been detected (Herbig, 1968; Jenkins et al., 1973) among some 18 molecules sought. CO is by far the most abundant in ζ Oph, as it is in other locations in the Galaxy, where some two-dozen interstellar molecules have been found by radio techniques. Yet in ζ Oph, CO accounts for less than 0.1% of the carbon expected on the basis of solar-system abundances. This again supports the idea that the depleted elements are in dust grains.

The question arises as to how representative is the cloud in front of ζ Oph. Morton et al. (1973) quote data for four other reddened stars, and find qualitative agreement with the depletion pattern of ζ Oph. Wallerstein and Goldsmith (1974) examined Ti in 14 stars. In half of them, $\log \xi \simeq -2$, while in the other half, $\log \xi < -2$ (as in ζ Oph). Chaffee (1974) studied Na, Ca, Fe, and Ti in three stars in Perseus. He found that $\log \xi \simeq 0$ for Na, $= -3.3$ to -2.9 for Ca, < -1 to -0.3 for Fe, and $= -2.3$ to -3.3 for Ti. These results agree roughly with Figure 2.

If the elements are in dust grains, there should be correlations between the depletion factors ξ and the properties of the elements. First, assume that the observed depletion is explained entirely by the nonequilibrium-accretion hypothesis. This depends on the charge on the grains (see above). If there are negatively charged grains present, positive ions would accrete on them only, as they would be completely repelled by positively charged grains. Light ions would deplete more rapidly than heavy ones. But the observations show that ^{7}Li and ^{23}Na are depleted very little, while ^{40}Ca, ^{48}Ti, ^{55}Mn, and ^{56}Fe are depleted more — often by substantial factors. Hence, negatively charged grains cannot be present if accretion is the sole explanation. On the other hand, if negatively charged grains were missing, one could not explain the fact that the three most depleted elements (Ca, Al, Ti) are ionized in the interstellar medium, for positively charged grains repel such ions completely. We therefore conclude that nonequilibrium accretion is not the sole explanation of the observed depletion factors. (This does not, however, say that this process does not play a role. In fact, even the original statement of the hypothesis by van de Hulst required some additional process to form condensation nuclei, as we have seen in Section 1.)

Next, consider the equilibrium hypothesis, according to which ξ should be given by equation (1) for a fixed value of T. In Figure 2, I have plotted ξ against T_c(m) for the 17 elements for which there is both abundance data and a value of T_c (Table 1). (It was assumed that T_c for Li equals that for Na and K.) Figure 2 differs from Field (1974) because T_c calculated for p = 1000 dynes cm^{-2}, rather than for p = 100, has been used. These generally differ by 100 K or so. A correlation seems evident in that, starting with S ($\xi \simeq 1$) at T_c = 700 K, there is a monotonic decrease of ξ with T_c, for Li, K, Na, Mn, Mg, and Ca, with the single exception of Si. The upper limits for B, Be, Fe, Ti, and Al are consistent with this trend. The fact that there is any correlation at all tends to support the equilibrium hypothesis. Moreover, the rapid drop between $T_c \sim 1000$ K and $T_c \sim 1800$ K is readily explicable in terms of the exponential dependence of ξ on T_c in equation (1), but is not easily explained otherwise. As expected, however, the data cannot be interpreted in terms of a single value of T. The observed (small) depletion of the alkali metals and the significant depletion of Mn suggest that sometimes T $\sim$ 1000 K. But if this were always so, application of equation (1) to Ca, for which T_c = 1650 K and ΔG = 160 kcal mole^{-1}, would yield log ξ = -13 — completely at variance with the observed value of -3.69. It seems quite reasonable that the freeze-out temperature ranges up to, perhaps, 1500 K, some gas-phase refractories being provided to the medium by those nebulae where T = 1500 K. If T were much lower than 1000 K in many cases, one would be at a loss to explain the near-normal abundance of S and Li, while if T were often much above 1500 K, no condensation would occur at all. Hence, a distribution of freeze-out temperatures between 1000 and 1500 K seems required by Figure 2 if equilibrium condensation in fact occurs.

4. NONEQUILIBRIUM CONDENSATION

We have shown that the nonequilibrium model alone, applied to oxygen-rich nebulae, cannot account for the observed depletion factors, while the equilibrium model has two points in its favor. It predicts observed condensates, and it predicts that ξ should drop sharply with T_c above T (Figure 2), as observed.

On the other hand, an oxygen-rich equilibrium model is not completely successful either. For one thing, it can explain at most about 1/3 of the observed extinction (Field, 1974; Greenberg, 1974, this volume). We are, apparently, forced to use the abundance elements O, C, N, and Ne (Table 2), which, up to this point in the discussion, have not been accounted for. All these elements are volatile, and we note that O, C, and N are observed to be depleted (Figure 2). While there is no data on Ne, another noble gas, Ar, is also depleted. The answer of van de Hulst (1949) to this problem would be straightforward: that these elements are present as ices, condensed on the grains already provided as nuclei. Indeed, Field (1974) showed that if the unaccounted-for O, C, and N are used in this manner, one can just account for the observed extinction in ζ Oph. But this solution is deceptively simple and requires further analysis.

Let us ask where this condensation might have taken place. If it occurred in oxygen-rich nebulae under equilibrium conditions, one would expect to form methane and ammonia hydrates (Table 2), but to do so would require freeze-out temperatures below 90 K to condense C and below 30 K to condense Ar. Such low temperatures would have depleted S, Li, K, and Na much more than observed. One also questions how a dust grain fragile enough to contain Ar could be accelerated away from a star without being destroyed.

Equilibrium condensation in carbon-rich nebulae would immediately account for at least part of the observed depletion of C, for graphite is among the first elements to condense, and graphite actually accounts for about 60% of the C in ζ Oph. Although this process probably takes place, it cannot account for all the carbon since $\xi(C) = 0.05$ in ζ Oph, and in any event, very low temperatures would be required to deplete O (as H_2O) and Ar, so the problem mentioned above would again be encountered.

Therefore, we seem to be forced to nonequilibrium accretion in space, which, as we have seen, is proportional to $(1 + 2.5 Z)A^{-1/2}$ for negatively charged grains. Since Li^+ has the largest value for this quantity, but is not much depleted, one conjectures that only positively charged grains, which repel ions like Li^+, are present. This leads to an immediate clarification, since of the four volatile elements observed to be depleted (O, C, N, and Ar), three (O, N, and Ar) are neutral in space because they have ionization potentials greater than hydrogen's. Thus, they alone would accrete on positively charged grains, as required. Furthermore, silicates are calculated to be positively charged.

Again, however, this solution is not completely satisfactory, for a number of reasons. The first, that the expected form of O (H_2O) is observed in only about $1/10$ the expected amount, can be explained by UV processing (Shulman, 1970; Donn and Jackson, 1970; Greenberg, 1972, 1973a,b); Khare and Sagan, 1973; Field, 1974). The second is a problem with carbon, which is ionized in space. On the one hand, C^+ will be repelled by silicates and, hence, will not accrete on them. However, we know that graphite is present, and graphite is calculated to be negatively charged. C^+ will be attracted to it, and it may well migrate to a growth edge and be incorporated into the graphite crystal (Field, 1974), thereby explaining the large amount of graphite observed in spite of the small number of carbon-rich systems where graphite nuclei could have formed under equilibrium conditions. But the presence of negatively charged grains, such as graphite, seems to be contradicted by the small depletion of Li. Here, the only way out appears to be to follow Aannestad (1972), who argued that binding energies of foreign atoms to graphite are much lower than to silicates, so they do not adhere to it. This suggestion is supported by the calculations of Gilra (1972), who finds that the observed 2200-Å feature can be explained in detail only if the graphite particles are virtually free of foreign materials.

Up to this point, the picture that emerges is one of negatively charged graphite grains, which accrete C^+ but not O, N, and Ne, and positively charged silicate grains, which accrete O, N, and Ne but not C^+, thus forming a mantle that is processed further by UV. This picture does not answer a number of questions. First, it does not account for the approximately 35% of the C that is neither graphite nor gaseous, according to the observations. One possibility (Field, 1974) is that as the "ice" mantles grow on the silicate grains, the photoelectric-ejection processes that keep them positively charged are inhibited, and the core-mantle grains become negatively charged, attracting C^+. The problem here is that the depletion factor of C calculated for the non-graphite component is $0.05/0.35 = 0.14$, and one would expect the depletion factor of Li^+ to be even smaller under these conditions, while the observed value is 0.55. Perhaps LiH, the molecule that forms from the H atoms on the surface, is more likely than CH to be ejected on formation (Field, 1974). The other depletion factors, referring to atoms heavier than C, may not be seriously in conflict with this possibility.

Another possibility is that in some oxygen-rich nebulae, carbon reacts through the Fischer-Tropsch process at about 400 K to form a large variety of hydrocarbons (which include some N), as postulated by Studier, Hayatsu, and Anders (1968) to account for the formation in the solar nebula of the organic compounds found in carbonaceous chondrites (see also Anders, 1971; Anders, Hayatsu, and Studier, 1973). This possibility is particularly attractive, because a result would be that some of the silicate grains would bear mantles containing organic molecules, among which are those seen in interstellar space by radio techniques (Anders, 1973). For this to account for the 86% of the nongraphite C missing from the gas phase, at least this fraction of interstellar matter would have to cycle through oxygen-rich nebulae that cooled below 400 K. Obviously, S, Li, K, and Na would be totally condensed in these cases, so ξ would be < 0.14 for these cases, contrary to observation. It seems possible that both processes operate to some extent to reduce ξ (C). Hence, the best model at present may be one that includes pure graphite grains and silicate-core grains with mantles of volatile elements. These mantles are mostly composed of compounds of N, O, and some C accreted in interstellar space, but are sometimes composed of compounds of C and N (and some O) formed by the Fischer-Tropsch process in a small fraction of the particularly cool nebulae. Both types of mantles would also have some Ne.

5. DISCUSSION

If this model is at all close to the truth, it has implications for the frequency of occurrence of condensation sites. Consider Ca, for which ξ is 1/5000 in ζ Oph and of the order of 1/1000 in Perseus (Chaffee, 1974). Some of this may be due to non-equilibrium accretion in interstellar space, but even under favorable conditions (negatively charged grains, doubly ionized Ca), Ca depletes no faster than C (Mészáros, 1972). Since some of the depletion of C is due to equilibrium processing through carbon-rich nebulae, we conclude that the nonequilibrium accretion of Ca results in $\xi_N > 0.05$ [$= \xi(C)$]. Suppose a fraction f of the interstellar medium has never been at an equilibrium-condensation site; its depletion is $\xi = \xi_N$. For the fraction 1-f of the medium that has been at such sites, the value of ξ is $\xi_E \xi_N$. Hence, the observed depletion is $\xi = f\xi_N + (1 - f)\xi_E \xi_N$. If we conservatively adopt $\xi < 0.001$ and $\xi_E = 0$, it follows that

$$f < \frac{0.001}{0.05} = 0.02 \quad , \tag{2}$$

so that less than 2% of the interstellar medium has never been at a condensation site.

Now consider the cycling of the interstellar medium through such sites. The present mass of the medium is called M and the original mass is called M_o. If, averaged over a large number of events, a fraction s of the gas involved (and the accompanying dust) goes permanently into stars and is lost, and a fraction 1-s is returned to the interstellar gas with all calcium condensed into grains (obviously the most favorable case), one can show (Appendix B) that

$$f = \left(\frac{M}{M_o} \right)^{(1-s)/s} \quad . \tag{3}$$

In this calculation, supernova injection of fresh Ca is ignored; since it will be gaseous, it can only place an even greater strain on the condensation process.

From equations (2) and (3), together with the fact that $M/M_o \simeq 0.02$, we conclude that 1-s > 0.5, so that more than half of the gaseous Ca must be returned in condensed form.

Now consider the implications of this for the astronomical nature of the site. If the site is a stellar atmosphere, it is required that a large fraction of all stellar matter is ejected back into space on a short time scale. In a Salpeter mass function, only 1/3 of the matter goes into stars with masses > 1 M$_\odot$, which have a chance of ejecting a significant amount of matter back into the medium. Truran and Cameron (1971) estimate that perhaps 30% of the mass of such stars is ejected back into the medium. Since less than 100% of the Ca involved will be in condensed form, $1-s < 0.1$, contrary to the above requirement. A further problem with stellar ejecta is that deuterium is destroyed in stars, so the fraction of primordial D that remains is just the fraction f that has never been in stars in this model. If D were primordial, this would require that the primordial D/H is >50 times the present one (1.4×10^{-5}; Rogerson and York, 1973), or 7×10^{-4}. This would require $\rho_0 = 4 \times 10^{-32}$ g cm^{-3} in a standard big-bang model (Wagoner, 1973), corresponding to $\Omega \sim 0.01$, only marginally consistent with the observed mass in galaxies (Shapiro, 1971). According to Truran and Cameron (1971), D/H has fallen by a factor of only 6, up to the present. If D is produced in supernovae (Colgate, 1973) this (weak) argument would be effectively countered (but see Reeves, 1973, for arguments against SN production).

If, on the other hand, condensation occurs in nebulae, we require that more matter goes into the nebulae and is subsequently ejected (in condensed form) than goes into the star. While the requirement of effectively complete condensation is hard to meet, the concept that a massive nebula is ejected has already been advocated for other reasons (Cameron and Pine, 1973). Since this phenomenon should occur for low-mass as well as high-mass stars, it does not seem impossible that $1-s > 0.5$, averaged over all star-formation events. If so, this would constitute an argument that nebulae such as the solar nebulae are commonplace and, therefore, that planets beyond the solar system are also.

ACKNOWLEDGMENTS

This work was supported in part by NSF under GP-36194X. It benefitted from discussions with A. Cameron, E. Purcell, E. Anders, J. Greenberg, G. Herbig, and E. Jenkins.

REFERENCES

Aannestad, P. A., 1972. Ph.D. Thesis, University of California at Berkeley.

Aannestad, P. A., and Purcell, E. M., 1973. Ann. Rev. Astron. Astrophys. 11, 309.

Anders, E., 1971. Ann. Rev. Astron. Astrophys. 9, 1.

Anders, E., Hayatsu, R., and Studier, M. H., 1973. Science 182, 781.

Anders, E., 1973. Molecules in the Galactic Environment, ed. by M. A. Gordon and L. Snyder (Wiley, New York), p. 429.

Bless, R. C., and Savage, B. D., 1972. Astrophys. Journ. 171, 293.

Cameron, A. G. W., 1973. Space Sci. Rev. 15, 121.

Cameron, A. G. W., Colgate, S. A., and Grossman, L., 1973. Nature 243, 204.

Cameron, A. G. W., and Pine, M. R., 1973. Icarus 18, 377.

Chaffee, F. H., Jr., 1974. Astrophys. Journ., in press.

Colgate, S. A., 1973. Astrophys. Journ. (Lett.) 181, L53.

Donn, B., and Jackson, W. M., 1970. Bull. Amer. Astron. Soc. 2, 309.

Dorschner, J., 1968. Astron. Nach. 290, 171.

Feuerbacher, B., Willis, R. F., and Fitton, B., 1973. Astrophys. Journ. 181, 101.

Field, G. B., 1974. Astrophys. Journ., to be published.

Gilman, R. C., 1969. Astrophys. Journ. (Lett.) 155, L185.

Gilra, D. P., 1971. Nature 229, 237.

Gilra, D. P., 1972. The Scientific Results from the Orbiting Astronomical Observatory OAO-2, ed. by A. D. Code (NASA SP-310), p. 295.

Greenberg, J. M., 1972. Mem. Soc. Roy. Sci. Liege 3, 425.

Greenberg, J. M., 1973a. On the Origin of the Solar System, (Centre National de la Recherche Scientifique, Paris), p. 135.

Greenberg, J. M., 1973b. Molecules in the Galactic Environment, ed. by M. A. Gordon and L. E. Snyder (Wiley, New York), p. 93.

Grossman, L., 1972. Geochim. Cosmochim. Acta. 36, 597.

Herbig, G. H., 1968. Zs. f. Astrophys. 68, 243.

Herbig, G., 1970. <u>Evolution Stellaire Avant la Séquence Principale</u>, Mem. Soc. Roy.
    Sci. Liege <u>19</u>, 13.

Hoyle, F., and Wickramasinghe, N. C., 1962. Mon. Not. Roy. Astron. Soc. <u>124</u>,
    417.

Jenkins, E. B., Drake, J. F., Morton, D. C., Rogerson, J. B., Spitzer, L., and
    York, D. G., 1973. Astrophys. Journ. (Lett.) <u>181</u>, L122.

Khare, B. N., and Sagan, C., 1973. <u>Molecules in the Galactic Environment</u>, ed. by
    M. A. Gordon and L. E. Snyder (Wiley, New York), p. 399.

Lewis, J. S., 1972. Icarus <u>16</u>, 241.

Mészáros, P., 1972. Astrophys. Journ. <u>177</u>, 79.

Morton, D. C., Drake, J. F., Jenkins, E. B., Rogerson, J. B., Spitzer, L., and
    York, D. G., 1973. Astrophys. Journ. (Lett.) <u>181</u>, L103.

O'Keefe, J. A., 1939. Astrophys. Journ. <u>90</u>, 294.

Purcell, E. M., 1969. Astrophys. Journ. <u>158</u>, 433.

Purcell, E. M., 1973, private communication.

Reeves, H., 1973. 13th Intl. Conf. on Cosmic Rays, Denver, Colorado, in press.

Rogerson, J. B., and York, D. G., 1973. Astrophys. Journ. (Lett.) <u>186</u>, L95.

Shapiro, S., 1971. Astron. Journ. <u>76</u>, 291.

Shulman, L. M., 1970. <u>Interstellar Gas Dynamics</u>, I.A.U. Symp. No. 39, ed. by
    H. J. Habing (Springer-Verlag, New York), p. 326.

Spitzer, L., 1948. Harvard College Observatory Centennial Symposia. Harvard Obs.
    Monograph No. 7.

Spitzer, L., Drake, J. F., Jenkins, E. B., Morton, D. C., Rogerson, J. B., and
    York, D. G., 1973. Astrophys. Journ. (Lett.) <u>181</u>, L110-115.

Stecher, J. P., and Donn, B., 1965. Astrophys. Journ. <u>142</u>, 1681.

Studier, M. H., Hayatsu, R., and Anders, E., 1968. Geochim. Cosmochim. Acta.
    <u>32</u>, 151.

Truran, J. W., and Cameron, A. G. W., 1971. Astrophys. Space Sci. <u>14</u>, 179.

van de Hulst, H. C., 1949. Rech. Astron. Obs. Utrecht <u>11</u>, (part 2), 41.

Wagoner, R. V., 1973. Astrophys. Journ. <u>179</u>, 343.

Wallerstein, G., and Goldsmith, D. W., 1974. Astrophys. Journ. <u>187</u>, 237.

Watson, W. D., 1972. Astrophys. Journ. 176, 103 and 271.

Whipple, F. L., 1946. Astrophys. Journ. 104, 1.

Whipple, F. L., 1950. Astrophys. Journ. 111, 375.

White, R. E., 1973. Astrophys. Journ. 183, 81.

Wickramasinghe, N. C., and Guillaume, C., 1965. Nature 207, 366.

TABLE 1.

Elemental condensation temperatures in oxygen-rich gas.

Temperature (°K)	Condensate	Name	Elements removed
1760	Al_2O_3	Corundum	Al
1650	$CaTiO_3$	Perovskite	Ti
1630	$Ca_2Al_2SiO_7$ $Ca_2MgSi_2O_7$	Melilite	Ca
1470	(Fe, Ni)	Metallic nickel–iron	Fe, Ni
1440	Mg_2SiO_4	Forsterite	Mg
1350	$MgSiO_3$	Enstatite	Si
1350	$BeAl_2O_4$		Be
1140	MnS	Alabandite	Mn
~1000	$(Na, K)AlSi_3O_8$	Alkali feldspar	Na, K
700	FeS	Troilite	S
700	$NaBO_2$		B
200	H_2O	Water	O
130	$NH_3 \cdot H_2O$	Ammonia hydrate	N
90[*]	$CH_4 \cdot XH_2O$	Methane hydrate	C
30	Ar (solid)	Argon (solid)	Ar
8	Ne (solid)	Neon (solid)	Ne

[*] Carbon may also condense as hydrocarbons at ~400 K (Anders, 1971).

TABLE 2.

Major condensates of the most abundant elements in oxygen-rich gas.[*]

Element	Log abundance, H = 12.00 (Cameron, 1973)	Condensates
O	8.83	Methane hydrate $(CH_4 \cdot XH_2O)$
C[†]	8.57	Methane hydrate $(CH_4 \cdot XH_2O)$
N	8.07	Ammonia hydrate $(NH_3 \cdot H_2O)$
Ne	8.03	Neon (solid)
Mg	7.53	Enstatite $(MgSiO_3)$
		Forsterite (Mg_2SiO_4)
Si	7.50	Enstatite $(MgSiO_3)$
		Forsterite (Mg_2SiO_4)
		Fayalite (Fe_2SiO_4)
Fe	7.42	Metallic nickel-iron (Ni, Fe)
		Troilite (FeS)
		Fayalite (Fe_2SiO_4)
		Magnetite (Fe_3O_4)
S	7.20	Troilite (FeS)

[*]After Grossman (1972) and Lewis (1972).

[†]See footnote to Table 1.

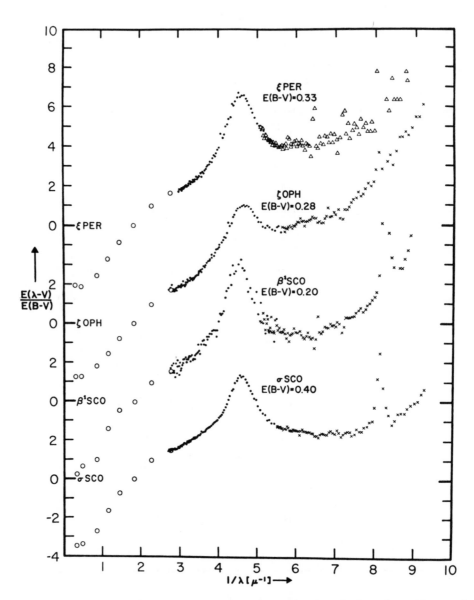

Figure 1. Normalized extinction curves based on OAO-2 and other data (Bless and
Savage, 1972).   The bump at $\lambda^{-1} = 4.5 \; \mu^{-1}$ is present in all of the ~200 stars
observed.   The magnitude of the bump in ζ Oph indicates that about 60% of
the carbon along the line of sight is in the form of small graphite flakes.
(Courtesy of the Astrophysical Journal.)

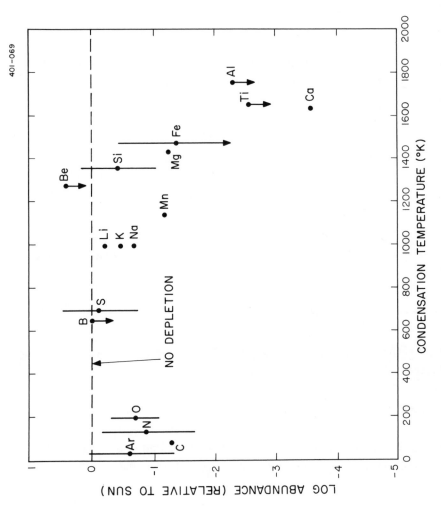

Figure 2. Abundances of various elements in the gas phase along the line of sight to ζ Oph, plotted logarithmically as a fraction of the corresponding abundance in the solar system. The condensation temperature, calculated from thermodynamics, is that temperature at which the element in question begins to condense in an oxygen-rich atmosphere at a pressure of 1000 dynes cm^{-2}. Normal abundance is indicated by the horizontal dotted lines. Note that most elements are depleted in the gas phase by factors up to 3000 (for calcium).

APPENDIX A

VAPOR DENSITY

Consider an element z, of which there are $n(z)$ atoms in molecule m. If the molar density (moles cm^{-3}) of the molecule is $s(m)$, that of the atom is

$$s(z) = n(z) \ s(m) \quad .$$

At high temperatures, the elements are in atomic form, with

$$s(z) = A(z) \ s(H + He) \equiv A(z)s \quad ,$$

where $A(z)$ is the relative abundance by number. As the temperature falls, various molecules m form, each with $n(z)$ atoms. The molar density of the molecules is given by

$$s(m) = \frac{s(z)}{n(z)} = \frac{A(z)}{n(z)} \ s(H + He) \quad , \tag{A1}$$

but $s(z)$ does not change. Finally, $s(m)$ falls below the vapor density, and condensation to the solid begins. Then the molar density of m is given by

$$\log [s(m)T] = C(m) - \frac{\Delta G(m)}{2.3RT} \quad , \tag{A2}$$

rather than by equation (A1), where $C(m)$ is a constant and $\Delta G(m)$ is the free energy of the transition. The condensation temperature $T_c(m)$ is defined by the condition that the vapor is saturated:

$$\log \left[\frac{A(z)sT_c}{n(z)} \right] = C(m) - \frac{\Delta G(m)}{2.3RT_c(m)} \quad . \tag{A3}$$

Subtracting equation (A3) from equation (A2), we see that

$$\log \xi = \log \left[\frac{T_c(m)}{T} \right] - \frac{\Delta G(m)}{2.3R} \left[\frac{1}{T} - \frac{1}{T_c(m)} \right] \quad , \tag{A4}$$

where

$$\xi = \frac{s(z)}{A(z)s}$$

is the depletion factor in the gas phase. Hence $\log \xi = 0$ for $T > T_c(m)$, and it is given by equation (A4) for $T < T_c(m)$.

In many cases, the molecule that is stable in the solid phase does not correspond to those that are stable in the gas phase; a chemical reaction takes place on condensation. It can be shown, however, that equation (A4) is still correct.

APPENDIX B

GAS REMAINING AS A RESULT OF PROCESSING IN NEBULAE

We assume that the relative abundance of each heavy element, including both gas and dust, does not change. As explained in the text, this assumption, although not correct if new heavy nuclei are being produced in supernova explosions, nevertheless permits a lower limit on the fraction 1-s of the matter that is ejected as nebulae to be deduced from observation.

Let M be the total mass of a heavy element in the interstellar medium. It changes permanently by an amount dM as the result of incorporation into stars (processing through nebulae does not change M). Associated with the change dM is a change in the mass of the element in the gas phase, fdM.

On the other hand, the total mass in the gas phase is fM. Of the change in gas mass d(fM) associated with star formation, only a fraction s goes into stars. Hence, we have a second expression for the change in gas mass resulting from the permanent change in M, sd(fM). Hence,

$$sd(fM) = fdM \quad , \tag{B1}$$

to which the solution is

$$f = \left(\frac{M}{M_0}\right)^{(1-s)/s} \quad , \tag{B2}$$

as stated in the text.

INTERACTION OF GAS AND DUST IN THE INTERSTELLAR MEDIUM

W. D. Watson

University of Illinois, Urbana, Illinois

ABSTRACT

Physical processes involved in the interaction of interstellar gas and dust are discussed. It is still not understood whether particles heavier than helium can be returned to the gas when they hit grain surfaces. Ejection by ultraviolet radiation seems to be the most likely process. Interpretation of observations of the H_2 molecule by the Copernicus satellite indicates that H_2 is formed in grain surfaces at a rate that is in semiquantitative agreement with theoretical predictions.

1. INTRODUCTION

Observable effects on dust grains of their interaction with the interstellar gas include alteration of their extinction properties and the presence of at least one infrared absorption line. Carrasco, Strom, and Strom (1973) observe changes in reddening and polarization of starlight (as a function of distance from the edge of an interstellar cloud), which they attribute to condensation of gas particles onto the grains. The long-sought 3.1-μ line of frozen H_2O has apparently been observed (Gillett and Forrest, 1973; see also Woolf, 1973). Most likely, this ice results from condensation of the gas in interstellar clouds onto the refractory cores of interstellar grains; the cores are thought to be formed in the atmospheres of stars. These observations as well as other upper limits (Knacke, Cudaback, and Gaustad, 1969) indicate that no more than about 10% of the material in interstellar grains is H_2O ice (Hunter and Donn, 1971; Woolf, 1973).

Potentially observable effects on the gas of the interaction between interstellar gas and dust typically are more interesting than effects on the grains. These include the formation of H_2 and complex interstellar molecules, neutralization of ions in dense interstellar clouds, depletion of carbon, nitrogen, oxygen, and heavier elements by amounts that may reflect the nature of the grain—gas interaction, and heating of the interstellar gas.

Formation of the H_2 molecule by H atoms sticking to interstellar grains has been analyzed in detail by Hollenbach and Salpeter (1970, 1971) and Hollenbach, Werner, and Salpeter (1971). Experimental evidence is consistent with this theoretical work, and observations of the H_2 molecule in the interstellar gas are in semiquantitative agreement with the predicted formation efficiency (see Section 7). While it is generally agreed that interstellar H_2 forms on grains, the long-accepted viewpoint that more complex molecules are somehow formed on grain surfaces is under reconsideration at present. Recent work (Herbst and Klemperer, 1973; Watson, 1973a, b) has shown that most of the smaller molecules can also be produced through ion—molecule reactions in the gas phase. In fact, the HD molecule can be produced in the observed quantity only by gas reactions (Watson, 1973c), and certain others (e.g., DCN, HNC, HCO^+) are most likely produced in the gas phase. However, the basic gas-phase reactions, e.g.,

$$D^+ + H_2 \rightarrow HD + H \quad ,$$

$$CH^+ + H_2 \rightarrow CH_2^+ + H \qquad \text{(leading to CH)} \quad ,$$

$$O^+ + H_2 \rightarrow OH^+ + H \qquad \text{(leading to OH)} \quad ,$$

and

$$H_2^+ + H_2 \rightarrow H_3^+ + H \qquad \text{(leading to a number of molecules in dense clouds)} \quad ,$$

all require H_2, and thus the presence of grains is necessary for its formation. The chief obstacle for the formation of complex molecules on grains is understanding how they are returned to the gas (e.g., Watson and Salpeter, 1972a, b; designated hereafter as WS1 and WS2). For the very low fractional ionizations expected to occur in dense interstellar clouds, recombination of ions by sticking to a grain can be a major factor in reducing the electron density (Watson, 1973b).

In the interstellar gas, depletion of heavy elements relative to hydrogen in comparison with cosmic-abundance ratios has been suggested for some time (e.g., Spitzer, 1968; Field, Goldsmith, and Habing, 1969). Recent observations by the Copernicus satellite (Morton et al., 1973) indicate that the more abundant heavy elements (for example, carbon and oxygen) tend to be underabundant by an order of magnitude in interstellar clouds. Heavier elements typically are underabundant by a larger factor. Although this depletion is thought to be a result of incorporating the gas atoms into grains, it is unclear whether this occurs mainly when the grains are formed in stellar atmospheres or whether it is a result of freezing onto the grains in the interstellar clouds (see Field, 1974). In the latter case, atoms that are ionized in the interstellar gas (e.g., carbon) would be expected to exhibit different depletion from neutral atoms, since the grains almost certainly have a significant electric potential. No such difference is readily evident from the observations (Morton et al., 1973). More complicated effects that depend on the differences in surface chemistry for the various elements may also occur and be reflected by differences in depletion. Abundances of carbon and oxygen are of particular importance because collisional excitation of their fine-structure lines provides the chief cooling mechanism for interstellar clouds. Continued depletion of the elements with time may have a considerable effect on inter-stellar cloud evolution (Mészáros, 1972).

Photoemission from grains (Watson, 1972) and formation energy of H_2 molecules, as suggested by Copernicus observations (Spitzer and Cochran, 1973), may be major sources of energy input for maintaining observed temperatures of interstellar clouds. The first process is uncertain because photoemission efficiencies are sensitive to the composition of the outer layers of the grain, which are not known. Some recent models for interstellar clouds (Glassgold and Langer, 1973) suggest that this photo-emission heating fits the observational data better than do other heating mechanisms for interstellar clouds with molecular hydrogen. The formation energy of an H_2 molecule is normally expected to go into vibrational excitation that is radiated away. If, however, the observed wide lines from H_2 molecules are due to recombination with most of the formation energy becoming translational, the result is an appreciable energy input into the gas.

The chief purpose of this review is to summarize present knowledge of certain physical mechanisms involved in the above phenomena that result from the interaction of interstellar gas and grains. We first discuss certain basic ideas – the relevant astronomical time scales, the "equilibrium" balance between the gas and dust, and the possible types of binding for a gas particle to a grain surface (Section 2). Photo-electron emission and the resulting electric potentials of interstellar grains, as well as the heat input to the gas, are treated in Section 3. While even hydrogen atoms stick to grains for at least a short time (Section 4), the subsequent ejection of heavy atoms is highly uncertain from present knowledge of ejection mechanisms (Section 5). Whether or not they are subsequently ejected, atoms hitting dust grains are always converted into molecules irrespective of uncertainties in the surface physics (Section 6). Finally, information on grain processes can be deduced from recent astronomical observations, e.g., formation efficiency of H_2 (Section 7).   In conclusion, our present knowledge is summarized in relation to the basic astrophysical problem of the grain−gas processes (Section 8).

2. BASIC CONSIDERATIONS

The time required for a particular atom of thermal speed v to hit any grain of cross section σ_g and number density n_g in the interstellar gas is

$$t_a = \left[\left(\frac{n_g \sigma_g}{n_0}\right) vn_0\right]^{-1} \approx \frac{10^9}{n_0} \text{ years } .\tag{1}$$

Here we have assumed that the surface area of grains per hydrogen atom $n_g \sigma_g/n_0$ is $\approx 10^{-21}$ cm^2, as indicated by recent ultraviolet observations. We adopt the usual viewpoint that $n_g \sigma_g/n_0$ is approximately constant in all regions of the interstellar gas, independent of the total hydrogen density n_0.

For gas particles to condense out onto grains or for most of the gas atoms to be converted into molecules, t_a must be shorter than the time between cataclysmic events in an interstellar cloud that would disrupt these phenomena. For condensation, the time scale is that required for disruption of the cloud itself or for a sufficiently violent cloud−cloud collision, both of which are about 10^8 years. During such a collision or when the interstellar cloud decays to become part of the intercloud gas (temperature $\approx 10^4$ K), destruction of the frozen grain mantles is likely (Aannestad, 1973a, b). For molecule formation, typical cloud−cloud collisions occurring every $\approx 10^7$ years are adequate to destroy molecules. From equation (1), it is thus clear that between disruptive events for densities $n_0 \stackrel{>}{\sim} 100$ cm^{-3}, 1) condensation of the gas onto grains is possible if every particle that hits does stick, and 2) collisions with grains can possibly convert most atoms to molecules. Observationally, the presence of large fractional abundances of H_2 in interstellar clouds shows that essentially every H atom has hit a grain, and thus that t_a is somewhat shorter than the time between disruptive events.

For dense interstellar clouds − where most complex molecules are observed − the lifetime may be as short as the time required for the cloud to contract for star formation. A lower limit on the contraction time is the free-fall time for a cloud,

$$t_f = \frac{5 \times 10^7}{\sqrt{n_0}} \text{ years} \quad . \tag{2}$$

Again, it is clear that for the typical dense cloud ($n_0 \gtrsim 10^3$ cm^{-3}), $t_a < t_f$, and inter-action with the grains can have a major effect on the gas.

Another interesting time scale, especially with regard to molecule formation, is that for a particular grain to be hit by some gas particle of number density n_p,

$$t_g = (\sigma_g n_p v)^{-1} \approx \frac{10^5}{n_0} \quad \text{sec for hydrogen}$$

$$\approx \frac{10^8}{n_0} \quad \text{sec for a heavy atom (C, N, O)} \quad . \tag{3}$$

Whether gas particles remain on a grain depends on the rates for a number of possible ejection processes. Nonthermal processes will be discussed in Section 5. Here we consider only steady thermal evaporation (vapor pressure), since this is uncontroversial. A particle bound to a grain surface of temperature T_g with a binding energy D is ejected in a time

$$t_t \approx \nu_0^{-1} \exp\left(\frac{D}{kT_g}\right) \quad , \tag{4}$$

where ν_0 is lattice frequency. A criterion for condensation of the gas onto grains is that a gas particle spend more of its time on a grain than in the gas, or

$$t_t > t_a \quad . \tag{5}$$

The analogous criterion for molecule formation involving heavy atoms is that after a heavy atom sticks to a grain, it must remain <u>at least</u> until a second atom (hydrogen) can stick to the grain, or

$$t_t > t_g(H) \quad . \tag{6}$$

As a result of the exponential dependence of equations (5) and (6), D/kT_g is the quantity of chief importance. In Table 1, we give as a function of D/k the critical values of T_g below which the inequalities (5) and (6) are valid.

The binding of an atom to the surface of an interstellar grain may depend sensitively on the nature of the surface and the particular atom involved. Being highly reactive, atoms are likely to attach themselves to a surface with chemical or semichemical binding energies. This is especially likely for "active" surfaces such as graphite, silicates, oxides, and SiC but is less likely for the relatively inert molecular crystal surfaces such as ice. For the former group, a layer of chemically adsorbed atoms may form; the binding on top of this layer is likely to be weak. Experimental evidence exists (see WS1 and Section 6) in favor of weak binding for H atoms under such conditions, as well as on ice surfaces. In the absence of chemical effects, a particle will always be bound to the surface by van der Waal's forces (physical adsorption). Thus, a minimum value of D can be obtained from consideration of physical adsorption alone. For a particular surface, D depends mainly on the polarizability of the particle. From the available data on physical adsorption energies and known polarizabilities of atoms and molecules, $D/k \gtrsim 800$ K is a good estimate for all particles of interest except H, H_2, and He. The special case of hydrogen, whose binding is unusually weak, will be discussed in Section 6.

Temperatures of interstellar grains under normal conditions (i.e., not near hot stars) do depend on composition and the optical shielding. For likely compositions, calculated temperatures T_g are 10 to 15 K (Werner and Salpeter, 1969; Greenberg, 1971). "Dirty ice" mantles, which now seem likely to be present, produce temperatures in this range.

From Table 1, it is clear that the $T_g \approx 10$ to 15 K are below the critical temperatures for equations (5) and (6) to hold, even for the minimum binding ($D/k \approx 800$ K) to the surface. Hence, if only steady thermal evaporation is considered for heavy elements, condensation onto grains is expected and the minimum requirement for molecule formation is met.

3. PHOTOEMISSION FROM GRAINS: GRAIN CHARGE AND HEATING OF THE GAS

Yields for photoemission from solids (photoelectrons per absorbed photon) usually increase rapidly up to energies of about twice the work function of the material. Thus, the range of photon energies of most importance for photoemission in the interstellar medium is $\approx$ 10 to 13.6 eV, where the radiation spectrum is cut off by hydrogen absorption. For many materials, including silicates, SiC, oxides, and probably forms of irregular and impure graphite, the photoemission yield is near 0.1 (see Watson, 1973d, for some representative yields). However, H_2O has an ionization potential close to 13.6 eV, and thus there is little photoemission from ice for photons in the interstellar gas. In addition, the yield from small grains of radii equal to a few hundred angstroms will be enhanced by a factor of 2 to 3 owing to geometrical effects (Watson, 1973e).

3.1 Grain Charge

In the absence of appreciable photoemission, the charge on a grain is determined by the equilibrium between the sticking rate of positive and negative particles from the gas. If both have the same sticking probability, a net negative charge will result, due to the greater speed of electrons. When protons are the chief positive particles and the gas has temperature T, a potential

$$\frac{eV}{kT} \approx -2.5 \tag{7}$$

is produced (Spitzer, 1968). The effective cross section of a negative grain for sticking by a positive ion (e.g., C^+) is enhanced by a factor of $1 + |eV/kT|$ over the geometrical cross section.

When the photoemission yield y is near 0.1 and the galactic ultraviolet radiation is unshielded, positive grain charges are produced. For the best value of the ultraviolet flux, a grain is positive if

$$20 \exp{(-2.5\tau)} \frac{\overline{y}/0.1}{n_e} \stackrel{>}{\sim} 1 \quad .$$

Here, $n_e(cm^{-3})$ is the electron density, τ is the optical depth at visual wavelengths from the edge of a cloud, and $\bar{y}$ is an average of the yield in the range 10 to 13.6 eV. Sticking of positive ions from the gas to a positive grain is then retarded by a factor exp (eV/kT) (Spitzer, 1968). In a previous paper (Watson, 1972), we have presented the voltages on an interstellar grain as a function of electron density assuming a yield y = 0.1. Positive grains with voltages up to 1 V result.

If depletion of heavy elements does occur from the gas in clouds, some differences are therefore expected between charged and neutral atoms. None is readily evident (Morton et al., 1973).

3.2 Heating of the Gas in Interstellar Clouds

Photoelectrons emitted from the grains typically have a kinetic-energy range ≈ 1 to 3 eV, which is large compared to the thermal energy kT $\approx$ 0.01 eV of the gas. Hence, a heating of the gas results, which can be expressed approximately as $10^{-14} n_0$ eV sec^{-1} for the assumptions of Section 3.1 and for no shielding of the galactic radiation (see Watson, 1972, for additional details). This heating rate is comparable to that for postulated low-energy cosmic or x-rays and is much greater than that for the observed sources of heating (high-energy cosmic rays). We have shown elsewhere (Watson, 1972) that the heating from photoemission can lead to interstellar cloud temperatures in reasonable agreement with observation, 30 to 80 K. This conclusion depends on the exact values of the relevant parameters − grain density and composition, and carbon abundance.

4. STICKING OF GAS PARTICLES TO INTERSTELLAR GRAINS

The question whether neutral gas particles will stick (i.e., become thermalized) to an interstellar grain with appreciable probability appears to be settled as a result of calculations by Hollenbach and Salpeter (1970) and further experimental work (Day, 1973; Marenco et al., 1972). Even hydrogen atoms stick with a probability of about 1/3 for interstellar cloud temperatures when only the weakest binding to the surface (van der Waal's) is assumed. Heavier neutral atoms (such as N and O) and molecules stick at essentially every collision.

In addition to the factors discussed in Section 3 that alter the sticking of a positive ion, its sticking may also be influenced appreciably by the electron recombination.
When the ion nears the surface, an electron from the grain probably recombines with
it while the ion is still outside the surface. Analysis of the details of this process
indicates that the sticking of positive ions to negative grains might be low (WS1).

5. NONSTEADY EVAPORATION MECHANISMS

5.1 Photoejection

Ejection by photons of a particle that is part of the lattice of a grain is negligible
even for astrophysical time scales. However, if the particle is only weakly bound to
the surface (physically or weak-chemically adsorbed), excitation of the particle may
cause its ejection (WS1). To a first approximation, the surface can be ignored in the
absorption of radiation by a weakly bound particle on the surface. The absorption can
result in exciting the particle onto a repulsive part of the potential curve for the upper
electronic state. Dissociation may then occur in the same manner as the dissociation
of molecules. Alternatively, the electronically excited state may decay into an unbound
state, so that the particle leaves the surface. If the binding of a particle to the surface
is weak, and if the particle is different from the material composing the surface,
it is likely that a fraction near unity of photons absorbed by such a particle causes its
ejection. For strong binding to the surface (chemical adsorption), photoejection of
particles is expected to be inefficient, in agreement with limited experimental data
(see WS1). It is convenient to define an efficiency for photon ejection of adsorbed
particles,

$$\epsilon = \frac{\text{rate at which a given adsorbed particle is photoejected}}{\text{rate at which particle is excited by uv radiation in interstellar gas}} \quad .$$

Thus, $\epsilon \gtrsim 0.01$ for strongly adsorbed particles, but we suggest $\epsilon \approx 1$ for weakly
adsorbed particles.

The only experiment to determine ϵ in the laboratory for weakly bound particles
has recently been performed by Greenberg (1973) for CS_2, CO_2, O_2, CO, C_6H_6, C_4H_{10},
N_2, CH_4, H_2O, CH_3OH, and NH_3 on a cold quartz substrate with radiation of energy

$\gtrsim 6$ eV photon^{-1}. Although photoejection was observed for all these molecules, interpretation of the experimental yield in terms of an efficiency ϵ involves some added uncertainty. Values $0.01 \gtrsim \epsilon \gtrsim 1$ are suggested. It is expected that ϵ will tend to increase for the higher frequency radiation (6 to 13.6 eV) that is also present in the interstellar medium.

Until further experimental studies have been performed for various surfaces, photoejection from interstellar grains must be employed with caution. In any event, it is likely that the density of starlight in dense interstellar clouds is sufficiently low that photoejection is not important. Additional factors must be considered.

5.2 Ejection by Cosmic Rays

The processes that might potentially be effective (WS1) are   1) evaporation due to heating of a spot around the path of the cosmic ray, and 2) for small grains, heating of the entire grain to a high temperature for a short period. The latter process is significant because grain temperatures are well below the Debye temperatures of solids. Hence, the specific heat of the grain is low, and high temperatures can be achieved for modest input energy. Duley (1973) has considered the analogous temperature pulses that result from absorption of photons by interstellar grains.

From the analysis of WS1, cosmic rays probably cannot evaporate significant quantities of gas from the grains, even if the cosmic-ray flux is as large as advocated for the ionization and heating of the interstellar gas (see Dalgarno and McCray, 1972). This conclusion does depend on gas density and the binding D of a particle to the grain surface. Recent observations strongly indicate that the interstellar cosmic-ray flux in clouds is much lower than has been suggested for heating and ionization (see O'Donnell and Watson, 1974).

5.3 Sputtering

For interstellar cloud temperatures, the energy (kT $\approx$ 0.01 eV) of the gas is not sufficient for a gas atom to knock particles from the surface of a grain. The number of higher energy, nonthermal particles is negligible.

When an interstellar cloud has decayed and become part of the "intercloud gas," the grain collides with gas particles that have kinetic energies of about 1 eV. Calculations (Aannestad, 1973a) indicate that ice mantles acquired by grains in clouds will be destroyed in the intercloud phase. A grain will also encounter energetic gas particles during cloud−cloud collisions. Under favorable conditions, which occur approximately every 10^8 years, the hot gas that results from the collision may evaporate ice from grains by sputtering (Aannestad, 1973b). Sputtering under these conditions does not, however, prevent accumulations of gas particles onto grains on time scales less than 10^8 years − i.e., the time required for an atom to stick to a grain.

6. MOLECULE FORMATION ON INTERSTELLAR GRAINS

In the foregoing sections, it was established that 1) gas particles stick to a grain with a probability of order unity per collision, 2) particles of interest, other than H or H_2, are not ejected by steady thermal evaporation before other gas particles with which they can react stick to the grain, and 3) after molecule formation, it is uncertain (except for H_2) whether the molecule is ejected from the surface of the grain.

The remaining two physical processes of importance are the mobility of particles on the grain surface and the ejection of the molecules in the formation process (see WS1). Because the surface potential is not smooth, but has "ridges" associated with the positions of particles from which the lattice is composed, a barrier between lattice sites exists that inhibits the motion of a particle. The particle moves either by thermal hopping over the surface barrier or by quantum mechanical tunneling through the barrier. Thermal hopping is a simple classical process, and its rate can be obtained with confidence as a function of E_B/kT_g, where E_B is the height of the surface potential barrier. Quantum mechanical tunneling is thought to be the dominant cause of atomic hydrogen motion on an inert surface, such as ice, to which it is weakly bound. Hollenbach and Salpeter (1971) calculate that, given reasonable assumptions about surface irregularities, an H atom will traverse the entire surface of an ice grain in about 10^{-6} sec as a result of tunneling. Quantum mechanical tunneling may also be important for heavier atoms (C, N, O) that are physically adsorbed. Both types of mobilities depend on the height of the surface barriers. If the atom is strongly bound to the surface (chemisorbed), it will not be mobile, but it will also not be ejected from the surface. One or more layers of material will form on the surface. On top of these layers, which are probably ice or other inert molecular crystals, the binding for atoms will be relatively weak.

Formation of H_2 is somewhat special, owing to the probable weak binding of atomic hydrogen to a regular grain surface. For a surface that only physically adsorbs atomic hydrogen, a limited number of surface sites of enhanced binding D are required. Otherwise, steady thermal evaporation will eject the H atom from the grain before a second atom, with which it can react, adheres. Irregularities in the surface due to growth edges of the grain, pitting of the surface by cosmic rays or other energetic particles, and the presence of impurity atoms are expected to produce sufficient enhanced binding D for formation of H_2. In the formation process, the H_2 molecule is expected to convert a sufficient fraction of its recombination energy into translational energy to overcome the binding to the surface. If the binding of atomic hydrogen to the entire surface is stronger than physical adsorption that is due to chemical or semi-chemical bonds, formation of H_2 is only improved. Hollenbach and Salpeter (1970) predict that H_2 molecules are ejected from the surface in high rotational states ($J \approx 8$), in high vibrational states ($v \approx 12$), and with translational energies of about 0.20 eV. These predictions are in agreement with suggestions that the highly excited rotations of H_2 molecules observed by Copernicus may in part be created in these states (Spitzer and Cochran, 1973). Uncertainties in the nature of the grain surface and in the calculations are such that the work of Hollenbach and Salpeter (1970) does not seem to exclude the possibility that as much as 3 eV of the recombination energy is converted into translational energy, as suggested by Spitzer and Cochran (1973). An experimental investigation (Marenco et al., 1972) has established 1) that H_2 can be formed from H atoms on an ice surface under laboratory conditions, 2) the efficiency of formation is consistent with predictions, 3) the energy of formation given to the surface is probably small, so that the H_2 molecule is ejected during formation with most of its energy in rotational, vibrational, and translational modes, and 4) atomic hydrogen is weakly attached (physically adsorbed) to an ice surface.

If weakly bound, atoms other than hydrogen will be sufficiently mobile on a grain surface to find other atoms and react, even if their mobility is due only to thermal hopping. Strongly bound, immobile heavy atoms will either react with mobile H atoms or will form a layer on top of which the binding is weak enough to allow molecule

formation. The inefficiency of photoejection for strongly bound particles ensures
that immobile particles are not ejected fast enough to prevent molecule formation.
From approximate calculations, it appears that, during the molecular recombination, some fraction of the resultant molecules are most likely ejected directly, and
some probably remain on the surface (WS1). Those remaining may be photoejected
if the starlight is not strongly shielded. In any case, starlight is well shielded in
dense clouds, and there is a clear problem involved in returning molecules to the gas.
For the less dense clouds, the time required for particles to stick to a grain is only
comparable with the cloud lifetime, and it is possible that the observed particles constitute the fraction ejected during recombination. Shocks (Aannestad, 1973b) and the
possibility that the starlight in dense clouds is greater than generally assumed
(Grasdalen, Strom, and Strom, 1973) offer some hope for ejecting molecules from
grains in dense clouds.

7. ASTROPHYSICAL EVIDENCE

Although the widespread observations of H_2 by Copernicus suggest an efficient
formation process, the uncertainties in the structure of interstellar clouds (especially
density and mass distribution) prevent an accurate determination of the exact efficiency.
From the methods of Hollenbach, Werner, and Salpeter (1971), constant density
models for interstellar clouds can be constructed that yield column densities of H and
H_2 as a function of

$$\frac{\text{formation rate of } H_2 \text{ per H atom}}{\text{photodestruction rate of } H_2 \text{ in unshielded starlight}} = \frac{n_g \sigma_g v \epsilon}{2\Gamma} \quad .$$

Here, n_g and σ_g are the number density and cross section of grains, v the velocity of
H atoms, ϵ the efficiency of H_2 formation when an H atom hits a grain, and Γ the
photodissociation rate of H_2 unshielded starlight. For the observed H and H_2 in the
direction of ζ Oph, we obtain $n_g \sigma_g v \epsilon / 2\Gamma \approx 1 \times 10^{-4}$. If the accepted values
$n_g \sigma_g / n_0 \approx 10^{-21}$ cm^{-2} and $\Gamma \approx 10^{-10}$ sec^{-1} are adopted, $n_0 \epsilon_0 \approx 200$. Values for the
total hydrogen density n_0 can be estimated from various arguments, but they are
unreliable. Densities $n_0 \approx 100$ cm^{-3} are usually thought to occur in those clouds that
cause somewhat more reddening than do standard clouds. Hence, $\epsilon \approx 1$, in excellent
agreement with predictions of Hollenbach and Salpeter (1971). The likely range for
n_0 is such that probably $1 \gtrsim \epsilon \gtrsim 0.10$.

8. SUMMARY

Formation of heavy interstellar molecules on grains and the related question of the depletion of heavy gas atoms remain among the most important unsolved problems of the gas−grain interaction in H I clouds. Recent observations by Copernicus and one laboratory experiment can be interpreted as strongly indicating that H_2 is formed on grains at near the rate predicted by Hollenbach and Salpeter (1971).

For heavy atoms, it is still not understood what prevents their complete freezing out onto grains, unless it is simply a matter of time scales. Some molecules formed on grains are ejected in the formation. Some may be ejected by photoejection when starlight is present. It is thus possible that a fraction of the heavy atoms hitting a grain are returned to the gas and that the remainder freeze to form mantles.

REFERENCES

Aannestad, P. A., 1973a. In Interstellar Dust and Related Topics, Proc. IAU Symp. No. 52, ed. by J. M. Greenberg and H. C. van de Hulst (D. Reidel Publ. Co., Dordrecht-Holland), p. 341.

Aannestad, P. A., 1973b. Astrophys. Journ. Suppl. 25, 223.

Carrasco, L., Strom, S. E., and Strom, K. M., 1973. Astrophys. Journ. 182, 95.

Dalgarno, A., and McCray, R. A., 1972. Ann. Rev. Astron. Astrophys. 10, 375.

Day, K., 1973. In Interstellar Dust and Related Topics, Proc. IAU Symp. No. 52, ed. by J. M. Greenberg and H. C. van de Hulst (D. Reidel Publ. Co., Dordrecht-Holland), p. 311.

Duley, W. W., 1973. Nature (Phys. Sci.) 244, 57.

Field, G. B., 1974. Astrophys. Journ. 187, 453.

Field, G. B., Goldsmith, D. W., and Habing, H. J., 1969. Astrophys. Journ. (Lett.) 155, L149.

Gillett, F. C., and Forrest, W. J., 1973. Astrophys. Journ. 179, 483.

Glassgold, A. E., and Langer, W. D., 1973. In preparation.

Grasdalen, G. L., Strom, K. M., and Strom, S. E., 1973. Astrophys. Journ. (Lett.) 184, L53.

Greenberg, J. M., 1971. Astron. Astrophys. 12, 240.

Greenberg, L., 1973. In Interstellar Dust and Related Topics, Proc. IAU Symp. No. 52, ed. by J. M. Greenberg and H. C. van de Hulst (D. Reidel Publ. Co., Dordrecht-Holland), p. 413.

Herbst, E., and Klemperer, W., 1973. Astrophys. Journ. 185, 505.

Hollenbach, D., and Salpeter, E. E., 1970. Journ. Chem. Phys. 53, 79.

Hollenbach, D., and Salpeter, E. E., 1971, Astrophys. Journ. 163, 155.

Hollenbach, D., Werner, M. W., and Salpeter, E. E., 1971. Astrophys. Journ. 163, 165.

Hunter, C. E., and Donn, B., 1971. Astrophys. Journ. 167, 71.

Knacke, R. F., Cudaback, D. D., and Gaustad, J. E., 1969. Astrophys. Journ. 158, 151.

Marenco, G., Schutle, A., Scoles, G., and Tommasini, F., 1972. Journ. Vac. Sci. Tech. 9, 824.

Mészáros, P., 1972. Astrophys. Journ. 177, 79.

Morton, D. C., Drake, J. F., Jenkins, E. B., Rogerson, J. B., Spitzer, L., and York, D. G., 1973. Astrophys. Journ. (Lett.) 181, L103.

O'Donnell, E. J., and Watson, W. D., 1974. Astrophys. Journ., in press.

Spitzer, L., 1968. Diffuse Matter in Space (Interscience, New York), p. 124.

Spitzer, L., and Cochran, W. D., 1973. Astrophys. Journ. (Lett.) 186, L23.

Watson, W. D., 1972. Astrophys. Journ. 176, 103.

Watson, W. D., 1973a. Astrophys. Journ. (Lett.) 183, L17.

Watson, W. D., 1973b. Astrophys. Journ., in press.

Watson, W. D., 1973c. Astrophys. Journ. (Lett.) 182, L73.

Watson, W. D., 1973d. In Interstellar Dust and Related Topics, Proc. IAU Symp. No. 52, ed. by J. M. Greenberg and H. C. van de Hulst (D. Reidel Publ. Co., Dordrecht-Holland), p. 335.

Watson, W. D., 1973e. Journ. Opt. Soc. Amer. 63, 164.

Watson, W. D., and Salpeter, E. E., 1972a. Astrophys. Journ. 174, 321.

Watson, W. D., and Salpeter, E. E., 1972b. Astrophys. Journ. 175, 659.

Werner, M. W., and Salpeter, E. E., 1969. Mon. Not. Roy. Astron. Soc. 145, 249.

Woolf, N. J., 1973. In Interstellar Dust and Related Topics, Proc. IAU Symp. No. 52, ed. by J. M. Greenberg and H. C. van de Hulst (D. Reidel Publ. Co., Dordrecht-Holland), p. 485.

TABLE 1.

Critical temperatures of grains for condensation of the gas $T_g(5)$ and molecule formation $T_g(6)$ at a total number density $n_0 = 10 \text{ cm}^{-3}$. D/k is the binding energy of a particle to the surface.

D/k (°K)	$T_g(5)$	$T_g(6)$
500	9	13
800	14	21
1000	18	26
1500	27	39
2000	36	52
3500	63	—
6500	120	—

EFFECTS OF PARTICLE SHAPE ON VOLUME
AND MASS ESTIMATES OF INTERSTELLAR GRAINS

J. Mayo Greenberg

State University of New York at Albany and Dudley Observatory, Albany

and

Seung Soo Hong

State University of New York at Albany

ABSTRACT

Mass estimates of interstellar grain materials based on visual extinction characteristics are shown to be insensitive to shape and, so long as the wavelength dependence of extinction is defined well into the infrared, they are also insensitive to size distribution. Spheroidal particles are treated by an approximate analytical method. Spheres and cylinders (core mantle as well as homogeneous) are treated by exact methods.

1. INTRODUCTION

Most estimates of the mass densities of interstellar grains have been based on spherical models. This has been done primarily for convenience, because the introduction of nonsphericity appears to produce substantial complications. Although some estimates of the effects of nonsphericity (Greenberg, 1972, 1973; Greenberg and Hong, 1973) seem to indicate that the effects are small when $a/\lambda \simeq 1$ and are significant only when $a/\lambda \ll 1$, recent quantitative work on dust and cosmic abundance has made it necessary to make more refined calculations in order to justify these estimates clearly.

In this paper, we shall consider not only the effects of shape but also the effects of certain size distributions, and even some effects of inhomogeneity of grains, i.e., core-mantle particles. Although we shall not thoroughly exhaust all possibilities, the range of situations covered is sufficiently extensive that we believe the effects may be well estimated for many cases of interest. In the final section, we shall summarize the results of recent work on dust and cosmic abundance to show that including nonsphericity does not affect the conclusions. We shall also examine some effects on mass estimates of ice and silicates from infrared-absorption data at 3 and 10 μ, respectively.

Section 2 deals with particles that are small relative to the wavelength. Subsequent sections take up effects of shape, size distribution, and inhomogeneity when the particles are similar in size to the wavelength.

2. SMALL PARTICLES: $a/\lambda \ll 1$

Ordinary interstellar grains are generally less than $2a = 0.5$ μ in outer dimension so that observations in the infrared are such that $x = 2\pi a/\lambda \leq 0.5$ for $\lambda > 3$ μ. The Rayleigh approximation is valid for a wide range of complex indices of refraction as far as $x = 0.3$ within a few percent (van de Hulst, 1957), so there exists a significant range of particle sizes and infrared wavelengths in which we may apply it. Certainly the observations at 10 μ may be interpreted as implying x ≪ 1, and since a considerable segment of the grain-size population for any model lies in the a ≤ 0.15 μ range, we expect the Rayleigh approximation to be almost completely acceptable.

Thus, the result of Greenberg (1972) on the absorption of variously elongated and flattened spheroids of enstatite, ice, and metals should be indicative of the possible effects. It is shown there that the peak absorptivity per unit volume for the 3.07 μ ice band is increased at most 25% relative to spheres if one has infinitely thin disks and that very long needles increase the absorptivity by less than 2%. The integrated absorption in the ice band is increased by at most about 10 to 15% in the most extreme case of the thin disk.

The enstatite band as used is more complicated than the ice band, having three individual peaks. Thus, individual peak heights relative to the sphere may even be doubled in the case of the thin disk. In general, the effects of nonsphericity for enstatite are greater than those for ice. It is probably realistic to state that if the interstellar silicates are elongated or flattened by 2:1, then the mass estimates based on spheres may be perhaps 20% or so too high.

For metallic particles, the absorptivity depends very strongly indeed on shape. Thus, as absorbers, iron needles of about 25:1 ratio would be at least 10 times as efficient, per unit volume, as spheres.

For finite degrees of elongation or flattening, the sphere seems to provide the minimum absorptivity per unit volume. However, it is not possible to prove this to be completely general. As a matter of fact, it can be shown (Greenberg, unpublished) by a completely analytic calculation too long to give here, that in the limit of small eccentricity, the sphere may be either a maximum or a minimum depending on the values of m' and m''. For example, it turns out that, relative to prolate spheroids with axis aligned along the direction of polarization, the sphere gives minimum absorptivity in a rather small region around the origin in the m', m'' plane but is maximum everywhere else.

3. PARTICLES RESPONSIBLE FOR THE CLASSICAL "λ^{-1}" PORTION OF THE EXTINCTION CURVE: $x \simeq 1$

3.1 General

Given the amount of extinction (not absorption) over a distance D at wavelength λ, we can calculate the mass density for grains, all having the same chemical composition, from

$$\rho_d = \frac{\Delta m(\lambda) \; \Sigma V}{1.086D \; \Sigma C_{ext}(\lambda)} \; \overline{s} \; , \tag{1}$$

where ΣV is the total volume of the model grains, $\Sigma C_{ext}(\lambda)$ is the total extinction cross section of the model grains, and $\overline{s}$ is the specific gravity of the grain material. Henceforth, we shall define the ratio of volume to extinction as the volume-extinction factor, $V_C = \Sigma V / \Sigma C_{ext}(\lambda)$. When grains are represented by a single-sized sphere of radius $\overline{a}$, equation (1) becomes (Greenberg and Hong, 1973)

$$\rho_d = \frac{4}{3} \frac{\overline{a}}{\overline{Q}(\lambda)} \frac{\Delta m(\lambda)}{1.086D} \; \overline{s} \; , \tag{2}$$

where $\overline{Q}(\lambda)$ is an average extinction-efficiency factor at wavelength λ. We shall, in the following, let $\lambda = 5000$ Å and consider $\Delta m(\lambda)$ to be the visual extinction, Δm_v. Actually, λ_v is at about 5500 Å, so that although our relative comparisons are quite accurate, the values of $\Sigma V / \Sigma C_{ext}$ will be a few percent too low as applied to the visual.

In the following, we shall evaluate and compare the effects on the values V_C produced by consideration of size distribution, shape, and inhomogeneity, In particular, we shall determine whether some previously selected values of $\overline{a}$ and $\overline{Q}$ (Greenberg and Hong, 1973; Greenberg, 1974) are good representative values for the grain models considered. This will be discussed in the final section on applications.

3.2 Homogeneous Spherical Grains: Size Distributions

In this section, we shall determine what values of V_C are representative of several different grain materials if each of these sizes of grains is assumed to reproduce, at least qualitatively, the wavelength dependence of extinction in the visual region.

Before showing the Mie theory results, it is instructive to carry out an approximate analytical calculation. The extinction efficiency for nonabsorbing spheres of low refractive index is (van de Hulst, 1957)

$$\frac{C}{\pi a^2} = 2 - \frac{4}{\rho} \sin \rho + \frac{4}{\rho^2} (1 - \cos \rho) \quad , \tag{3}$$

where $\rho = 4\pi a \lambda^{-1}(m - 1)$, m = index of refraction. The ratio V_C for a particle size distribution n(a) is

$$V_C = \frac{4\pi/3 \int_0^\infty n(a) \, a^3 \, da}{\int_0^\infty n(a) \, C(a,\lambda) \, da} \quad . \tag{4}$$

For a Gaussian distribution

$$n(a) = A \exp \left[-\left(\frac{a}{a_0}\right)^2 \right] \quad , \tag{5}$$

equation (4) becomes (Greenberg, 1973)

$$V_C^{-1} = \frac{3\sqrt{\pi}}{a_0} \left\{ \frac{1}{4} - \frac{1}{2} \exp\left(\frac{-\zeta^2}{4}\right) + \zeta^{-2} \left[1 - \exp\left(\frac{-\zeta^2}{4}\right) \right] \right\} \quad , \tag{6}$$

where $\zeta = 4\pi a_0 \lambda^{-1} (m - 1)$.

For m = 1.33 ("ice"), it can be shown (Hayes et al., 1973) that $a_0 = 0.213 \, \mu$ produces an excellent representation of the wavelength dependence of extinction. Evaluating equation (6) at $\lambda^{-1} = 2 \, \mu^{-1}$, we obtain $V_C = 0.185$. We note that this result implies that the highly simplified form of equation (2) should be quite adequate for ices if one uses $\bar{a} = 0.2 \, \mu$ and $\bar{Q} = 1.5$ to give $V_C = 0.178$, as in Greenberg (1974).

Now we shall show, by the use of Mie theory, that this latter value of V_C is an excellent choice for ices, and further we shall derive values of V_C for other grain materials.

The exact Mie theory calculations have been applied to evaluating equation (4) with a size distribution of the form (Greenberg, 1966)

$$n(a) = A \exp\left[-5\left(\frac{a}{a_i}\right)^3\right] , \tag{7}$$

where several values of a_i have been used for each grain material to show how the appropriate value is selected to produce a fair representation of the visual range of the extinction curve.

Normalized extinction curves for silicate [$(Mg,Fe)SiO_3$], magnetite (Fe_3O_4), graphite (C), and silicon carbide (SiC) are given in Figure 1. Equivalent figures for ices are available in Greenberg (1968), in which it was shown that for this substance, $a_i = 0.5 \mu$ is a proper choice. For other substances, the choices are as underlined in Figure 1 and as specified in Table 1. The indices of refraction used are given in Table 2. It is to be noted that none of the silicon carbide curves are good extinction curves; therefore, we chose the least objectionable.

We have not exhausted size-distribution effects. Perhaps more monodisperse size distributions might modify the V_C factors, although the indicated comparison in Table 1 with the crude delta-function choices does not seem to produce substantial differences. This is under further investigation.

3.3 Shape Effects

It is well known that, given a total surface area, the volume of a particle is maximized by the spherical shape. The implication that the interstellar particles with a given extinction (proportional to their area) require less material if they are nonspherical than if they are spherical is, however, not justified. The reason for this is simply summarized by noting, for example, that if one were to take a set of spherical particles and elongate them, the shape of the wavelength dependence of extinction by the randomly oriented elongated particles would be qualitatively like that of smaller spheres. Thus, in order to obtain the same wavelength dependence of extinction, we should have started with large spheres (but smaller numbers) in order to maintain the same total extinction. Thus, the total volume of material turns out to be essentially shape-invariant because it is constrained not only by the amount of visual extinction but also by the wavelength dependence.

Although precise computational methods are available (Reilly, 1969) for the scattering by almost any smooth convex particle, we shall, for convenience, restrict ourselves here to consideration of infinite cylinders, and, for finite nonspherical particles, we shall use the analytical approximation for spheroids (Greenberg, 1960, 1968). It is of some interest to note that in Figure 5 of the 1960 reference, a comparison of the extinction per unit volume was made and only small effects of nonsphericity were indicated even when quite extreme cases were considered.

Consider spheroids with semirotational axis R, semitransverse axis T, and elongation e = R/T. The ray approximation for the extinction cross section is (Greenberg, 1968)

$$C_{ext} = 4 \, AB \, Re \left\{ \frac{1}{2} - \left(\frac{iT}{C\rho} \right) \exp \left(- \frac{C\rho}{T} \right) - \left(\frac{iT}{C\rho} \right)^2 \left[1 - \exp \frac{iC\rho}{T} \right] \right\} \; , \qquad (8)$$

where

$$A^2 = T^2 \, (e^2 \sin^2 \theta + \cos^2 \theta) \; ,$$

$$B^2 = T^2 \; ,$$

$$C^2 = e^2 T^2 / A^2 \; ,$$

θ = angle between the rotational axis and the direction of light propagation,

$\rho = (4\pi T/\lambda) \, (m-1), \; m = m' - im'' $.

If we limit ourselves to the two orthogonal particle orientations in which $\theta = 0$ or $\theta = \pi/2$, it is possible to demonstrate analytically that V_C is not a function of elongation, $V_C \neq V_C(e)$. We standardize the wavelength of evaluation of the quantity V_C by choosing, for all elongations and orientations, a value of ρ_v given by $Q(\rho_v) = Q_{max}/2$, where Q_{max} is defined at the first major resonance of equation (8).

For $m'' = 0$, $\theta = \pi/2$, equation (8) reduces to

$$Q^{\perp}_{ext} = \frac{C_{ext}}{\pi T^2} = 4e \, [0.5 - \rho^{-1} \sin \rho + \rho^{-2} (1 - \cos \rho)] \quad , \tag{9a}$$

which is, except for the factor e, exactly the same as equation (3). Evaluating $\partial Q_{ext}/\partial \rho = 0$ numerically, we get $\rho_{max} = 4.09$. This gives $Q_{max} = 4e \times 0.793$, and letting $Q(\rho_v) = Q_{max}/2 = 4e \times 0.397$ gives $\rho_v = 1.99$. Thus, evaluating ρ_v at $\lambda = 5000 \, \mathring{A}$ gives $T = 0.24 \, \mu$ for $m = 1.33$, and from

$$C_{ext} = 4 \, eT^2 \times 0.397$$
$$V = \left(\tfrac{4\pi}{3}\right)eT^3 \quad ,$$

we get $V_C = T/(3 \times 0.397) = 0.202$, which is independent of the elongation e.

For $\theta = 0$, we get

$$Q^{\|}_{ext} = 4 \, \{0.5 - (e\rho)^{-1} \sin (e\rho) + (e\rho)^{-2} [1 - \cos (e\rho)]\} \quad , \tag{9b}$$

and performing the same computations as for equation (9a), we again arrive at $V_C = 0.202$ independent of e.

As a matter of fact, it is clear from the nature of the approximation leading to equation (8) that the extinction efficiency curves for spheroids at parallel and perpendicular incidence are of the same functional form as those of equivalent spheres and, consequently, evaluating V_C as we have done at a standard point on the extinction curve, must always produce the same value as for spheres.

For oblique incidence, the evaluation of V_C does depend somewhat on the elongation. However, rather than examine all such cases, which would, in any event, be unrealistic for the interstellar grain problem, we have numerically averaged the cross section over all different angles of incidence,

$$\langle C_{ext} \rangle = \frac{2}{\pi} \int_{0}^{\pi/2} C_{ext} (\theta) \, d\theta \quad , \tag{10}$$

to obtain the values of V_C given in Table 3.

It appears that V_C is a very slowly varying function of particle elongation. It is quite interesting and important to note that contrary to the naive expectation that the sphere should maximize V_C, we find that by including the wavelength dependence of the extinction criterion, the V_C for spheres is lower than that for spheroids.

The computations leading to Table 3 are intended to show relative shape effects and are not necessarily the best values, depending as they do on a scattering approximation. Nevertheless, the value of V_C = 0.206 for the sphere shown in Table 3 is really quite good, differing by only about 15% from the rather elaborately derived value of V_C = 0.178 for a size distribution of spheres using Mie theory. This indicates that the method for choosing the ρ value for evaluation is quite reasonable (Figure 2).

Our exact calculations of V_C for particles of varying elongation are limited to a comparison between spheres and cylinders. Only particles with ice-like indices of refraction were considered. More calculations for a wider range of indices are in progress. Table 4 presents calculations of V_C for single-sized spheres and for single-sized but spinning (averaged over angle) infinite circular cylinders that again roughly produce the right kind of extinction in the visual (see Figure 3).

We do not expect a single-sized particle to produce a really good match to the observed extinction, but we see from Figure 3 that a sphere of size 0.20 μ and m = 1.33 - 0.05i with V_C = 0.188 is pretty good, as is the cylinder of size 0.10 μ and either m = 1.33 - 0.00i with V_C = 0.238 or m = 1.33 - 0.05i with V_C = 0.210. Thus, we see that going from the sphere to equivalent very long particles implies a change in V_C from, say, 0.188 to 0.210, between which the difference is about 12%.

3.4 Core-Mantle Cylinder

The treatment here is rather directly oriented to a specific set of grain models (Greenberg and Hong, 1973; Greenberg, 1974) involving size distributions of spinning core-mantle cylinders. The mass estimates of the core and mantle materials were, however, estimated on the basis of very simple concentric sphere models, and we shall examine the reality of this estimate. It is obvious that spinning cylinders whose radii are a_{cyl} act like spheres of radii $a_{sph} > a_{cyl}$ (Greenberg, 1968), and consequently the effective a's for the concentric sphere should be larger than those of the cylinders by some factor empirically determined to be of the order of 1.25. Greenberg (1968) demonstrated that the best choice of cylinder radii matching the observed extinction is about four-fifths that for spheres of the same material. This factor was not included in the referenced papers (Greenberg and Hong, 1973; Greenberg, 1974), although it should have been.

For concentric spheres, if we let the core and mantle radii be a_c and a_m, respectively, and their ratio be $\alpha = a_m/a_c$, we find that the space densities of core (ρ_c) and mantle (ρ_m) material relative to the hydrogen density are

$$\frac{\rho_c}{n_H m_H} = \text{(numerical factor)} \left[\frac{4}{3} \frac{a_c}{Q} \frac{1}{\alpha^2} \right] \bar{s}_c \quad , \tag{11a}$$

$$\frac{\rho_m}{n_H m_H} = \text{(numerical factor)} \left[\frac{4}{3} a_c \frac{\alpha(1-\alpha^{-3})}{Q} \right] \bar{s}_m \quad , \tag{11b}$$

where $\bar{s}_c$ and $\bar{s}_m$ are respective specific gravities and where the numerical factor depends on the ratio of interstellar extinction to hydrogen density. The factors in the square brackets are the respective volume extinction factors, which when compared with concentric spinning cylinders of equal core and mantle radii, should, as already explained, be multiplied by 5/4.

A group of silicate-core–ice-mantle cylinder models that quite closely conform to the visual-wavelength dependence of extinction is characterized by cores of radius $a_c = 0.08~\mu$ with mantles distributed in size according to the form

$$n(a_m) = A \exp \left[-5 \left(\frac{a_m - a_c}{a_i} \right)^3 \right] \quad ,$$

where $a_i = 0.12$, 0.14, and 0.16 μ. The average, or effective, single values of a_m have been shown to be given by $a_m - a_c \simeq 0.3 a_i$ (Greenberg, 1968). Thus, single mantle thicknesses are respectively 0.036, 0.042, and 0.048 μ, which give $a_m = 0.116$, 0.122, and 0.128 μ.

The numerical calculations are performed to obtain the extinction from the total cross sections of constant-elongation cylinders for orthogonal polarizations of the incident radiation,

$$C_E = \frac{2}{\pi} \int_{a_c}^{\infty} n(a_m) \, a_m^2 \, da_m \int_0^{\pi/2} Q_{ext}^E (a_c, a_m, X) \, dX \quad ,$$

$$C_H = \frac{2}{\pi} \int_{a_c}^{\infty} n(a_m) \, a_m^2 \, da_m \int_0^{\pi/2} Q_{ext}^H (a_c, a_m, X) \, dX \quad , \tag{12}$$

where X is the tilt angle of the cylinder axis with respect to the direction of incident radiation. We then find ΣC_{ext} to be (closely)

$$\Sigma C_{ext} = \frac{C_E + C_H}{2} \quad . \tag{13}$$

The cylinder core and mantle volumes are calculated in the same way as is extinction — as if the particles of all sizes have the same elongation and the cores are as long as the mantles. Thus, the volumes of core and mantle are

$$V^c = 2\pi a_c \, e \int_0^{\infty} a_m \, n(a_m) \, da_m \quad ,$$

$$V^m = 2\pi \, e \int_0^{\infty} a_m \, (a_m^2 - a_c^2) \, n(a_m) \, da_m \quad , \tag{14}$$

for elongation (length/diameter) e. With values of C_{ext} from equation (13) and the volumes for equation (14), we can derive the results shown in Table 5.

We see from Table 5 that, with proper treatment, the core-mantle cylinder volume extinction factors are fairly well represented by the highly simplified core-mantle spheres. It is quite likely, but not yet demonstrated, that exact Mie-type calculations for core-mantle spheres would similarly show that the shape effect is not significant. It is important to note here that the conclusions of Greenberg (1974) regarding the atom-depletion discrepancy are not at all changed by using the more precise values given in Table 5.

4. APPLICATIONS

In establishing cosmic-abundance constraints on grain models, the key factor is V_C because it determines directly how much material is needed to provide a given amount of extinction. It is therefore critical to examine whether the conclusions (Greenberg and Hong, 1973) based on values of $V_C = (4/3) (\overline{a/Q})$ obtained from the highly simplified equation (2) can be justified.

We present in Table 6 a compilation of atomic abundances required by various homogeneous grain models as computed in the above reference and as estimated on the basis of effects of size distribution (Section 3.2) and on nonsphericity as much as 5:1 or 1:5 (Section 3.3).

It appears in Table 6 that the atomic-abundance constraints based on values of V_C for either spheres or spheroids of each material would be essentially identical.

ACKNOWLEDGMENT

This work was supported in part by grant NGR 33-011-043 from the National Aeronautics and Space Administration.

REFERENCES

Greenberg, J. M., 1960. Journ. Appl. Phys. 31, 82.

Greenberg, J. M., 1966. In Spectral Classification and Multicolour Photometry, Proc. IAU Symp. No. 24, ed. by K. Loden, L. O. Loden, and U. Sinnerstad (Academic Press, New York), p. 291.

Greenberg, J. M., 1968. In Nebulae and Interstellar Matter, ed. by B. M. Middlehurst and L. H. Aller (Univ. Chicago Press, Chicago), Chapter 6, p. 221.

Greenberg, J. M., 1972. J. Colloid Interface Sci. 39, 513.

Greenberg, J. M., 1973. In Molecules in the Galactic Environment, ed. by M. A. Gordon and L. E. Snyder (John Wiley & Sons, New York), p. 93.

Greenberg, J. M., 1974. Astrophys. Journ. (Lett.), 189, L81.

Greenberg, J. M., and Hong, S. S., 1973. In Galactic Astronomy, Proc. IAU Symp. No. 60, ed. by F. J. Kerr and S. C. Simonson (D. Reidel Publ. Co., Dordrecht, Holland), in press.

Hayes, D. S., Mavko, G. E., Radick, R. R., Rex, K. H., and Greenberg, J. M., 1973. In Interstellar Dust and Related Topics, Proc. IAU Symp. No. 52, ed. by J. M. Greenberg and H. C. van de Hulst (D. Reidel Publ. Co., Dordrecht, Holland), p. 83.

Reilly, E. D., Jr., 1969. Ph.D. Thesis, Rensselaer Polytechnic Institute.

van de Hulst, H. C., 1957. Light Scattering by Small Particles (John Wiley & Sons, New York), p. 433.

TABLE 1.

Volume extinction factors V_C for several grain materials.

Grain material	Ice		Silicate	Magnetite	Graphite	Silicon carbide
m	1.33	1.33−0.05i		as in Table 2		
$a_i(\mu)$	0.5	0.5	0.25	0.12	0.10	0.12
V_C	0.173	0.175	0.0931	0.0489	0.0417	0.0507
V_C^*	0.178	0.178	0.0889	0.0513	0.0513	0.0444

*As obtained from equation (2), with $\bar{a} = 0.2$ μ for ice, $\bar{a} = 0.1$ μ for silicate, $\bar{a} = 0.05$ μ for magnetite, graphite, and silicon carbide, and $\bar{Q} = 1.5$ for the "dielectrics" (ice, silicate, silicon carbide) and $\bar{Q} = 1.3$ for the "metallics" (graphite, magnetite).

TABLE 2.

Index of refraction.

λ^{-1} (μ^{-1})	Silicate	Magnetite	Graphite	Silicon carbide
0.5	1.630−0.000i	3.95−1.46i	3.60−2.92i	2.50−0.00i
1.0	1.640−0.000i	2.24−0.42i	2.76−1.83i	2.58−0.00i
1.5	1.642−0.000i	2.55−0.45i	2.47−1.46i	2.60−0.00i
2.0	1.650−0.000i	2.45−0.65i	2.46−1.46i	2.65−0.00i
2.5	1.685−0.000i	2.40−0.81i	2.46−1.45i	2.75−0.00i
3.0	1.730−0.000i	2.22−0.90i	2.28−1.62i	2.80−0.00i
3.5	1.780−0.010i	2.15−0.94i	1.79−2.23i	3.00−0.06i

TABLE 3.

Volume extinction factors averaged over incidence angles for variously elongated dielectric spheroids. Upper values for m = 1.33 - 0.00i, lower values for m = 1.33 - 0.05i.

Shape	Oblate		Sphere		Prolate		
elongation	0.2	0.5	1.0	2.0	3.0	5.0	10.0
T	0.820	0.347	0.246	0.174	0.161	0.164	0.167
	0.719	0.308	0.209	0.154	0.143	0.143	0.146
Q_v	0.938	1.103	1.587	2.206	2.977	4.690	9.104
	0.835	0.976	1.347	1.951	2.648	4.174	8.093
V_C	0.233	0.210	0.202	0.210	0.216	0.233	0.244
	0.230	0.211	0.206	0.211	0.216	0.229	0.240

TABLE 4.

Volume extinction factors for spheres and cylinders. Upper
values for m = 1.33, lower values for m = 1.33 - 0.05i.

Shape	Sphere		Cylinder	
a	0.20	0.25	0.10	0.15
Q_V	1.225	1.925	0.661	1.054
	1.414	1.997	0.749	1.118
V_C	0.218	0.173	0.238	0.223
	0.188	0.167	0.210	0.211

TABLE 5.

Volume extinction factors for spinning core-mantle cylinders and for concentric spheres. [*]

	Spinning cylinder				Sphere	
a_c (μ)	0.08	0.08	0.08	a	1.5	1.6
a_i (μ)	0.12	0.14	0.16	a_i(equiv)	0.14	0.16
V_C^c	0.0614	0.0560	0.0511	V_C^c	0.0395	0.0347
V_C^m	0.0840	0.0947	0.104	V_C^m	0.0938	0.107

[*]Calculated from the approximate equation (11) with the additional factor 5/4, as explained in the text.

TABLE 6.

Atomic densities for spherical and spheroidal homogeneous grain models

$$n_H/(\Delta m_V/D) = 2.34 \times 10^{21} \text{ atoms cm}^{-2} \text{ mag}^{-1}.$$

Material[*]	V_C	$\times 10^{-4}$				
		[O]/[H]	[Si]/[H]	[(Mg, Fe)]/[H]	[Fe]/[H]	[C]/[H]
modified ice	0.178	1.43				0.78
	0.175(1.15)	1.62				0.88
orthopyroxene	0.089	1.95	0.65	0.65		
	0.093(1.15)	2.34	0.78	0.78		
olivine	0.089	1.84	0.92	1.84		
	0.093(1.15)	2.21	1.10	2.21		
magnetite	0.051	1.25			0.94	
	0.049(1.15)[†]	1.38			1.03	
iron	0.051				1.07	
	0.042[‡](1.15)[†]				1.61	
silicon carbide	0.044		0.89			0.89
	0.051(1.15)		1.19			1.19
carbon	0.051					2.22
	0.042(1.15)[†]					2.10
cosmic abundance		6.76	0.32	0.34	0.26	3.70

[*] For each material, the upper values of V_C are as used in Greenberg and Hong (1973), and the lower values are estimated on the basis of size distribution effects with a multiplying shape factor (in parentheses) for a nonsphericity of 5:1 or 1:5.

[†] The shape effect was derived for dielectrics. The values given here are uncertain in application to metals.

[‡] Assumed to be the same as graphite.

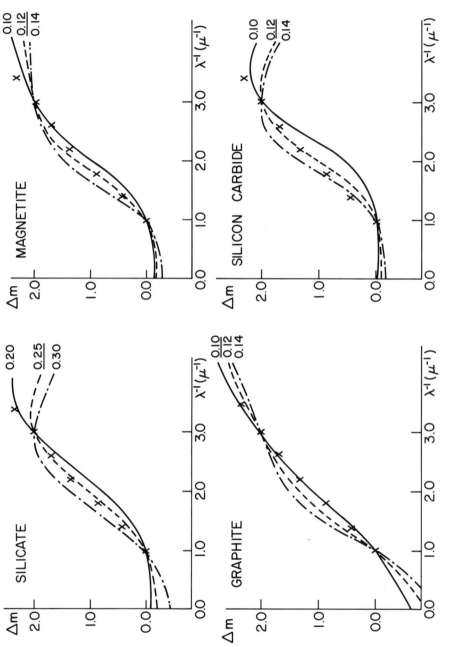

Figure 1. Normalized extinction curves for silicate, magnetite, graphite, and silicon carbide spheres with size distributions of the form $n(a) = A \exp[-5(a/a_i)^3]$. The cutoff sizes, a_i, are given in microns. Crosses represent the observational results.

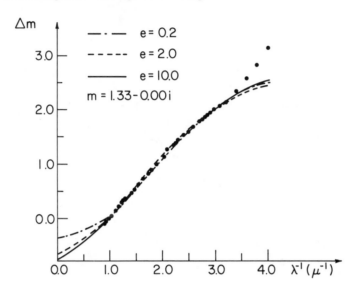

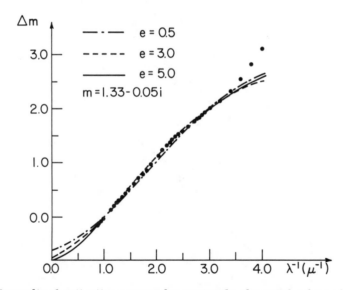

Figure 2. Normalized extinction curves for variously elongated spheroids calculated by a ray approximation. The method of choosing the particle size is as given in the text, where an appropriate value of $\rho = (4\pi a/\lambda) (m - 1)$ is assigned to correspond to the wavelength $\lambda = 5000$ Å [see equation (9a)], in order to reproduce the observed shape of the extinction curve. Dots represent the observational results.

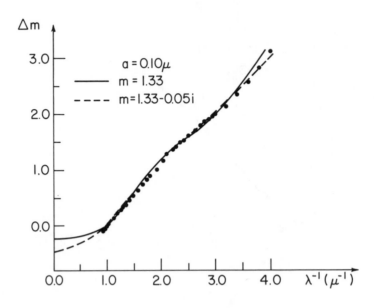

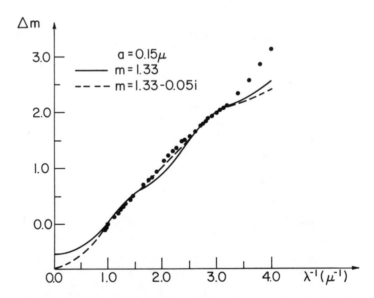

Figure 3a.   Normalized extinction curves for spheres calculated by Mie theory.   Dots
represent the observational results.

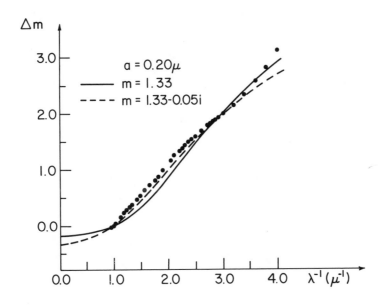

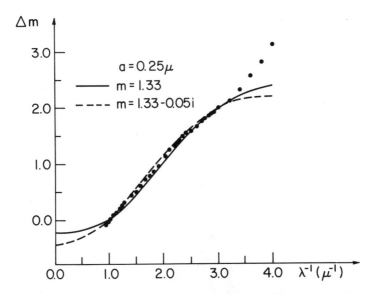

Figure 3b. Normalized extinction curves for single-sized spinning infinite cylinders.
Dots represent the observational results. Units are the same as in
Figure 1.

INTERSTELLAR GRAINS AS PINWHEELS

E. M. Purcell

Harvard University, Cambridge, Massachusetts

ABSTRACT

Contrary to the assumption usually made in theories of grain alignment, the rotational energy of an interstellar grain is likely to be very much greater than 3/2 kT, where T is the gas temperature, or the grain temperature, or any other temperature in the system. Any more or less permanent irregularity of the grain's surface with respect to accommodation coefficient, distribution of H-H recombination sites, or photoelectric emissivity will result in an unbalanced torque capable of spinning the grain up to high angular velocity. Such a grain is, in effect, the rotor of a heat engine. An unsymmetrical grain must eventually spin around either its axis of greatest inertia or its axis of least inertia. The relative preponderance of one outcome over the other, when magnetic torques and random impulses are taken into account, remains an unsolved problem about which a conjecture is offered. The consequences for magnetic alignment may be far-reaching.

It is still generally assumed that interstellar grains are aligned with respect to the interstellar magnetic field by the mechanism proposed long ago by Davis and Greenstein (1951). No other explanation of the polarization of starlight by interstellar matter has survived close examination. Accumulating observations, most notably the comprehensive study and compilation of Mathewson and Ford (1970), make it harder than ever to doubt that the polarization is a manifestation of the galactic magnetic field. On the other hand, recent studies of the Davis-Greenstein process (Jones and Spitzer, 1967; Purcell and Spitzer, 1971; Cugnon, 1971) have shown that adequate alignment of paramagnetic grains requires a magnetic field no weaker than a few times 10^{-5} G. It seems that one is obliged to assume either that the magnetic-field intensity over large regions is an order of magnitude greater than other lines of evidence and argument would indicate, or that the grains are ferromagnetic with rather special properties. Various grain compositions, not ferromagnetic, have been proposed to account for features of the interstellar extinction curve. All these proposals face the dilemma just stated. This very unsatisfactory state of affairs suggests that something essential may have been overlooked.

I suggest that we reconsider the rotation of grains in the interstellar environment. A grain bombarded by gas atoms in an H I cloud must, of course, obey the equipartition principle if the grain and the gas are at the same temperature T. The grain's average rotational kinetic energy will be 3/2 kT. On the other hand, if the grain is colder than the gas and if the accommodation coefficient — essentially, the sticking probability for a gas atom or molecule that hits the grain — is not zero, we would expect the mean rotational energy of the grain to be that appropriate to some intermediate temperature. And so it must be if the grain is sufficiently symmetrical. But it is easy to "design" a grain which, if colder than the gas, will be driven by the bombardment of gas atoms to rotate with a kinetic energy very much higher than $3/2 \, kT_{gas}$.

The old Crookes radiometer, a little paddle wheel of vanes shiny on one side and black on the other, pivoted inside a glass bulb, suggests the kind of heat engine I have in mind. It is unfortunate, pedagogically speaking, that the Crookes radiometer is actually driven in a complicated way by the residual gas in the bulb. Imagine that, instead, the bulb is totally evacuated and the pivots improved until the paddle wheel spins in response to light pressure when the radiometer is exposed to sunlight. The kinetic energy of the rotating paddle wheel is now enormously greater than it would be if it were immersed in, and in equilibrium with, a 6000 K radiation field, for then it would exhibit only a residual "Brownian rotation" appropriate to that temperature. So the wheel is really a heat engine, depending on the difference between the tempera- ture of the vanes and that of the radiation field to which they are exposed.

Similarly, the angular speed of an interstellar grain may greatly exceed that which random thermal excitation would cause, if the grain happens to embody in its shape or surface characteristics an element of rotational asymmetry of the "radiometer," or "anemometer," type, and if the requisite temperature difference exists.

Figure 1 shows an extremely simple design for an interstellar turbine, or "pinwheel" grain. This one is run not by light but by interaction with the interstellar gas. The only thing peculiar about this grain is its shape. Suppose the grain is colder than the ambient gas, the normal condition in an H I cloud. Suppose the accommodation coefficient is, say, 0.5 everywhere on the grain's surface. Obviously the two cavities, from which an entering gas atom is less likely to emerge "uncooled" than from an encounter with the convex surface, are effectively regions of higher accommodation coefficient. If, initially, the grain is not rotating, there will be, on the average, a net transfer of angular momentum from gas to grain, in the sense to cause clockwise rota- tion of the grain. The grain will spin up until its rotational speed is limited by gas friction, that is to say, by the fact that an advancing surface is more likely to be hit by an atom than is a retreating surface. At this limiting angular speed, the rotational kinetic energy of the grain is immense compared to kT_{gas}. One can easily show that it will be larger than kT_{gas} in the ratio, roughly, of grain mass to gas atom mass, M_g/m_H — typically 10^9. That holds for the extreme case (which actually prevails) in which the difference between T_{gas} and T_{grain} is comparable to or larger than T_{grain}. If the grain were kept hotter than the gas, it would of course spin in the opposite direction.

The factor 10^9 is so large that we may expect distinctly nonthermal behavior to result from even a small variation of accommodation coefficient over the surface of a wholly convex grain; an exotic shape is not really necessary. Imagine a grain that is not spherical, but a fairly compact lump – roughly brick shaped, for instance – with d as a typical dimension. Suppose the fractional variation of accommodation coefficient on the scale length d is δ. The expected rotational energy of such a grain after it has spun up to its limiting speed is, in order of magnitude, $\delta^2 (M_g/m_H) kT_{gas}$. Thus a variation of accommodation coefficient, or sticking probability, as small as 0.1% ($\delta = 10^{-3}$) could lead to rotational energy 1000 times thermal energy.

This is not the only way to drive a grain as a pinwheel. (Perhaps "anti-pinwheel" is the proper term for the example in Figure 1.) In fact, my attention was first attracted to the possibility of pinwheeling by a comment Professor Salpeter made on the manuscript of the paper by Spitzer and me on grain alignment. He pointed out that we had neglected to consider the excitation of grain rotation by recoil from the ejection of freshly formed H_2 molecules. According to the analysis of Hollenbach and Salpeter (1971), such molecules may be expected to depart with a modest fraction of the recombination energy, which makes quite a jolt. Indeed, these random jolts, from the point of view of Davis-Greenstein alignment theory, amount to an appreciable (but as it turns out, not decisive) increase in the apparent gas temperature.

I was struck with the thought that the jolts might not be in every sense random. For, again according to Hollenbach and Salpeter, recombinations $H + H \rightarrow H_2$ are supposed to take place at "active sites," of which there may be several but probably not an enormous number, on the grain's surface. These active sites are lattice defects or surface traps of one kind or another, where a migrating H atom, elsewhere held to the surface only by van der Waals forces, is bound until another H atom wanders up to react with it. If the location of such sites on a particular grain is a more or less permanent feature of that grain, the angular momentum imparted to the grain, averaged over many recombination events, will not be zero in grain coordinates. Merely the randomness in the distribution of ν sites over the grain may be expected to leave an unbalanced effect equivalent to something like the combined recoil from $\sqrt{\nu}$ sites. The grain will behave as if a rocket had been fastened to it, firing in a direction which, although unpredictable from grain to grain and varying rapidly in space as the grain rotates, is nevertheless fixed in "body coordinates," coordinates fastened to the grain.

The effect could be powerful. Consider an ensemble of grains in the shape of square platelets, of side b, thickness a, and mass M_g. Let there be ν active sites on each grain randomly distributed over its surface. Suppose that every H atom that hits the grain sticks and eventually leaves as part of a molecule ejected with energy E. Under these assumptions, the average rotational energy per grain in the ensemble will attain the order of magnitude $(M_g/m_H)\,(a/b)\,(E/\nu)$. For instance, if $b = 10^{-5}$ cm, $a = 10^{-6}$ cm, $E = 0.5$ eV and $\nu = 100$, the rotational kinetic energy is 5×10^4 eV, enormous compared to kT_{gas}. The rotational speed is around 10^9 revolutions $\sec^{-1}$.

Another mechanism capable of spinning up a grain is photoelectric emission. A 2-eV photoelectron takes away far more momentum than was brought by the photon that ejected it. That the photoelectric emissivity should vary over the surface of a grain is not merely plausible — the contrary is unthinkable. That the pattern of variation should be a sufficiently permanent feature of a given grain is much more doubtful. With a promise to face this question presently, let me assume that it is permanent. Then we can use the estimates of the rate of emission of photoelectrons already provided by Professor Watson (1974, this volume) to estimate the efficacy of the photoelectric pinwheel drive. If I assume that the emission of 2-eV electrons from a grain's surface occurs at the rate of 10^6 cm^{-2} $\sec^{-1}$, and that the gas has a density of 10 atoms cm^{-3} and a temperature of 100 K, I find that a grain of dimension d will acquire a rotational angular velocity of the order of 10^5 δ/d rad $\sec^{-1}$, where δ is here a measure of the unbalance in the distribution of photoemissivity over the grain. That is, $\delta = 10^{-3}$ would apply if one electron in a thousand was emitted from a site at the end of the grain, in a direction that would produce maximum angular impulse, while the remainder of the emission occurred with perfect isotropy. With that value of δ, for instance, which does not seem unreasonably high, a grain of size 10^{-5} cm would acquire an angular speed of 10^7 rad $\sec^{-1}$. Its kinetic energy, although not as large as in our example of the recombination pinwheel, is still 10^3 to 10^4 times kT_{gas}.

The three pinwheel mechanisms just described are summarized in Figure 2. A fourth might, in principle, be added — the direct radiometer action of starlight on optically unsymmetrical grains. But such an effect is insignificant for grains small compared to the wavelength of the light.

I have avoided until now two important questions. First, does not the rotation of the grain, carrying the "rocket" around with it, result in averaging out the rocket thrust so that no steady angular acceleration occurs after all? The answer is "no," because, in general, the vector representing the rocket torque in body coordinates will have a finite projection on the angular velocity vector, and, at least in the case where the sign of the residual torque is such as to increase the angular velocity, this will occur until finally the grain is rotating at its limiting speed around one of its principal axes of inertia.

A second question concerns the assumed permanence of the irregularity responsible for the rocket effect, be it irregularity in the distribution of accommodation coefficient, of recombination sites, or of photoelectric emissivity. The assumption is less plausible than one might at first think, because the time for a grain to spin up to its limiting speed is the same as the characteristic time τ_c for all grain–gas dynamical interactions, typically 10^5 to 10^6 years, in low-density clouds. If, through some process, a grain's surface were wholly altered in a time very much shorter than this, a given "rocket" would not last long enough to produce the effect I have described. The time τ_c is simply the time required for the grain to be struck by its own mass in gas atoms. Now the shortest time in which a grain could acquire a whole new surface is τ_c times the ratio of surface atoms to all atoms, or something like 10^{-3} to 10^{-2} τ_c. And if the rocket is capable of spinning the grain up to a rotational energy of 10^6 kT_{gas}, as we have seen in some instances, then, even in times as short as 10^{-2} τ_c, the kinetic energy could reach 100 kT_{gas}. This cursory examination, I admit, does not settle all questions of rocket lifetime, but I believe suprathermal rotation of interstellar grains is likely to prove the rule rather than the exception.

How does this affect the orientation of grains by the interstellar magnetic field? If a paramagnetic grain has attained a suprathermal rotational speed around a principal axis of inertia, the Davis-Greenstein interaction works to perfection. That is, the torque that arises because of the imaginary part of the magnetic susceptibility χ'', evaluated at the angular frequency of rotation ω_{rot}, will inexorably bring the axis of rotation into line with the magnetic field. In the ordinary case of a thermally rotating grain, the alignment is never complete, because of the random fluctuation in rotation caused by gas atoms hitting the grain. That is why, loosely speaking, so strong a magnetic field is needed to produce enough alignment to account for the observed polarization. But now we can expect eventual <u>complete</u> alignment, even for a much weaker field, given sufficient time. However, the high rate of spin maintained by the rocket does not make the alignment evolve more rapidly, since the angular momentum and the magnetic torque will have increased in the same ratio. The question of rocket lifetime remains, therefore, critical. The high rate of spin could conceivably make the alignment evolve somewhat more slowly, if ω_{rot} exceeds the frequency range within which the imaginary part of the magnetic susceptibility is proportional to ω. This is not likely to happen for $\omega_{rot} < 10^9 \text{ sec}^{-1}$.

The crucial question now emerges: About which of its principal axes of inertia is the pinwheel grain spinning? Think of the grain as a rectangular block of dimension $a > b > c$. The greatest moment of inertia, I_3, is that about the axis parallel to the c dimension; let I_1 be the moment of inertia about the axis parallel to the a dimension. In the absence of the magnetic field, stable rocket-driven rotation about either of these axes is possible; rotation about the intermediate axis is unstable. Including the rocket torque as a vector $\vec{N}$, fixed in body coordinates, and the gas friction as a torque determined by the angular velocity, Euler's equations for the motion in body coordinates of the angular velocity vector $\vec{\Omega}$ are

$$
\left.
\begin{aligned}
\dot{\Omega}_1 + \Omega_2\Omega_3(I_2 - I_3)/I_1 = N_1/I_1 - R\Omega_1 \\[2ex]
\dot{\Omega}_2 + \Omega_3\Omega_1(I_3 - I_1)/I_2 = N_2/I_2 - R\Omega_2 \\[2ex]
\dot{\Omega}_3 + \Omega_1\Omega_2(I_1 - I_2)/I_3 = N_3/I_3 - R\Omega_3
\end{aligned}
\right\} \quad . \tag{1}
$$

The following question, it turns out, is easily answered: If we start with the grains at rest $(\vec{\Omega} = 0)$ and turn on the rocket, causing constant torque components N_1, N_2, N_3 so that equation (1) henceforth applies, about which axis, axis 1 or axis 3, will the ensuing rotation eventually stabilize? A simple and exact criterion can be stated.[*] Unfortunately, this is not the problem we need to solve, for the rocket torque $\vec{N}$, over a short time scale, is exceedingly feeble compared to the fluctuating sum of purely random impulses. The development described by equation (1) has to evolve, not from

---

[*] One can show that axis 3 wins if and only if $N_3^2 I_1(I_3 - I_2) > N_1^2 I_3(I_2 - I_1)$.

$\vec{\Omega} = 0$ but from the "Brownian rotation" characteristic of thermal equilibrium. Nevertheless, only two asymptotic solutions are possible; some fraction P_3 of the grains will eventually spin around axis 3, and a fraction $P_1 = 1 - P_3$ will spin around axis 1. I do not know how to treat the passage from a stochastic to a deterministic behavior so as to predict confidently, for given $\vec{N}$ and given ratios of the moments of inertia, the value of P_3. Nor am I quite confident that introducing the magnetic torques into the dynamics would not influence the choice decisively. I regard this as a major unsolved problem.

The best I can offer now is little better than a "hand-waving" argument. Ignore the magnetic field and any rocket torques, and consider an ensemble of freely rotating asymmetric grains in thermal equilibrium with the gas atoms that occasionally bombard them. (Remember that whatever motion the grain is executing can't change in much less than 10^5 years, or about 10^{16} revolutions!) As viewed from axes fixed in the grain, the angular momentum $\vec{J}$ precesses either around axis 3 or axis 1. For a distribution of similar grains in statistical equilibrium, the fraction in each of the two classes can be deduced from the condition of equipartition of kinetic energy among the rotational degrees of freedom. One finds that W_3, the fraction of grains whose angular momentum vector is precessing around axis 3 (this being, as before, the axis parallel to the c dimension for our rectangular grain of dimensions $a > b > c$) is given by

$$W_3 = \frac{2}{\pi} \tan^{-1} \sqrt{\frac{b^2 - c^2}{a^2 - b^2}} \quad . \tag{2}$$

I now suggest that if a grain in this group carries a rocket, its fate will be to spin up around axis 3, for it is just the rocket torque component with respect to that axis that is not completely averaged out by the rotation. The spin-up will proceed directly if $\vec{N} \cdot \vec{J} > 0$. I'm not sure what happens if $\vec{N} \cdot \vec{J} < 0$. I suspect that an eventual spin-up around axis 3 in the opposite sense will result, but the crossover may be tricky. Let me assume that is the case merely to show the kind of conclusion we might reach if we could really solve the problem. If it is so, we can identify W_3 with P_3. Now we introduce the hitherto-neglected magnetic field $\underset{\sim}{B}$, which slowly lines up the already rapidly spinning grains, so that $\underset{\sim}{J}$ is finally parallel to $\underset{\sim}{B}$. In interesting contrast to the ordinary

Davis-Greenstein process, the sense of the resulting polarization of starlight would depend on the shape of the grains. For thin square plates, or disks, with $a \simeq b \gg c$, the alignment would be qualitatively like that expected in the ordinary Davis-Greenstein case but quantitatively more complete. The wide dimension of the grain would be perpendicular to the magnetic field. That would be consistent with the observed sense of polarization. Prolate grains, on the other hand, with $a \gg b \approx c$, would be mostly aligned with the long dimension along the field direction, according to this argument. That would produce the opposite polarization.

We see the tantalizing possibility of discovering whether the interstellar polarizing particles are needles or flakes, as well as accounting for the polarization without invoking an inordinately strong magnetic field. But this possibility will only be realized if someone solves the difficult dynamical — and also statistical mechanical — problem I have just posed: With what probability will axis 3 win?

Whatever its implications for magnetic alignment, the rapid rotation of grains induced by one or more of the pinwheel mechanisms might have other interesting consequences. I'll mention some that I have looked at without turning up anything easily observable. As Watson has emphasized, a grain is likely to acquire charge equivalent to a potential of one or two tenths of a volt. The rotating charged grain is a magnetic dipole; its axis of rotation will precess in the interstellar magnetic field. Here the drastic increase in rotational speed brings in nothing new, because the ratio of magnetic moment to angular momentum, which determines the precession frequency, is still the same. Martin (1971) has already shown that the precession rate is high enough to wash out any grain alignment except alignment with respect to the local magnetic field, and his conclusions remain in force.

More intriguing is the possibility of radiation from rapidly rotating charged grains, a possibility that had been raised in a somewhat different connection by Hoyle and Wickramasinghe (1967). One must expect the surface of a charged grain to be very nearly equipotential, so high are the field strengths at the surface and so long the times of interest. Nevertheless, a rotating grain of irregular shape can be, in effect, a rotating electric dipole if its center of mass and its center of charge do not coincide — as they would not for a pear-shaped grain, for instance. A displacement of the center

of charge as great as 1 or 2% of the grain "radius" is not unreasonable, I have concluded from a study of some exactly soluble shapes. Let us imagine a cloud of grains
of 10^{-5}-cm radius, charged to 0.3 V and rotating at the high rate we estimated for
the recombination-driven pinwheel, 10^9 revolutions sec^{-1}. Suppose that 10^9 Hz is
also the spread in rotation frequency. Is the 30-cm radiation from such a cloud
observable? The answer is that if the column density of grains is 10^{10} cm^{-2}, which
would produce a magnitude or two of visual extinction, the apparent temperature of the
cloud at 10^9 Hz is only 10^{-4} deg. It may be worth remarking that the damping of the
grain's rotation by its electromagnetic radiation is utterly negligible, even in this
rather extreme example.

An obvious consequence of high rotational speed is mechanical stress. An object
of size a, density ρ, and angular velocity ω is subjected to internal stresses of magnitude roughly $\rho a^2 \omega^2$. In our example above, if we assume density 1, this amounts to
3×10^9 dynes cm^{-2}. Such a stress would be supported by a compact crystal, but a
looser structure might well be torn apart. And a long time is available for plastic flow.
It seems possible that the morphology of the grains, or of some class of grains, may
be influenced by stresses from suprathermal rotation.

The mechanisms for driving the grain in rotation would be expected to produce
high translational kinetic energy as well. I have not been able to think of any interesting
consequences of this. The resulting speeds will still be well below atomic speeds.

In summary, I believe that suprathermal rotation must be not merely a possible
condition, but the usual condition of interstellar grains – perhaps excepting dark
molecular clouds. If "pinwheeling" does occur generally, it must play a role in grain
orientation, whether with the result that I have conjectured or with some other result.
In either case, one cannot cling to the ordinary Davis-Greenstein process unless one
can refute the argument that "pinwheeling" must be common. This is admittedly an
awkward position to arrive at, and I regret that I am unable to resolve it by a convincing solution of the dynamical problem. But perhaps these remarks will stimulate
others to try.

REFERENCES

Cugnon, P., 1971. Astron. Astrophys. 21, 85.

Davis, L., Jr., and Greenstein, J. L., 1951. Astrophys. Journ. 114, 206.

Hollenbach, D., and Salpeter, E. E., 1971. Astrophys. Journ. 163, 155.

Hoyle, F., and Wickramasinghe, N. C., 1967. Nature 214, 969.

Jones, R. V., and Spitzer, L., Jr., 1967. Astrophys. Journ. 147, 943.

Martin, P. G., 1971. Mon. Not. Roy. Astron. Soc. 153, 279.

Mathewson, D. S., and Ford, V. L., 1970. Mem. Roy. Astron. Soc., 74, 139.

Purcell, E. M., and Spitzer, L., Jr., 1971. Astrophys. Journ. 167, 31.

Figure 1. A simple interstellar turbine. Bombardment by atoms from the surrounding gas will cause this grain to rotate clockwise. The accommodation coefficient is 0.5 over all the surface, but an atom that strikes in a concavity is more likely to hit again before returning to the gas. If the gas were cold and the grain hot, the grain would rotate counterclockwise.

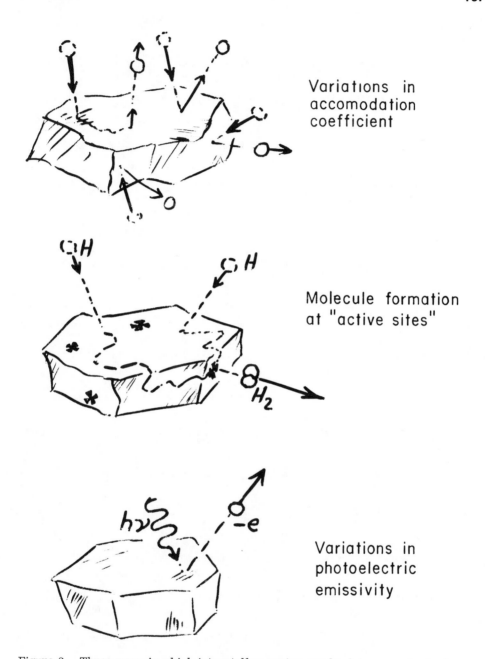

Variations in accomodation coefficient

Molecule formation at "active sites"

Variations in photoelectric emissivity

Figure 2.   Three ways in which interstellar grains can be driven as pinwheels.

COMETARY DEBRIS

R. E. McCrosky

Center for Astrophysics
Harvard College Observatory and Smithsonian Astrophysical Observatory
Cambridge, Massachusetts

ABSTRACT

Problems of ablation mechanisms of cometary nuclei are discussed in the light of some specific observations of comets and meteors. Estimates of the mass in the Geminid meteor stream are given. The outbursts of Comet P/Tuttle-Giacobini-Kresak are compared with those of the more distant comet P/Schwassmann-Wachmann 1. A formal solution of heat shock effects in comets near perihelion is given as an upper limit of the efficacy of this process for cometary disruption.

1. INTRODUCTION

Whipple's icy-conglomerate model of the comet nucleus proposes that these objects are seldom larger than some tens of kilometers in diameter and are composed of various ices and solid particles. We are virtually certain now, from ultraviolet spectra of comets, that water ice is a common constituent and suspect that ammonia and methane are also present. The near-identity of orbits between some meteor streams and particular comets is evidence of the solid material in the dirty-ice comet.

Much of our knowledge of comets is derived directly from studies of the gaseous and solid debris, and the crux of Whipple's argument favoring his comet model lies in the physical mechanism for producing the debris. In particular, the relative proportions of ice and dust and the rate of vaporization of the ices control the mass loss. Whipple's early papers dealt with the regime of moderate mass loss, where the outflowing gases are capable of carrying away the solids by rather gentle drag forces. It is the purpose of this paper to discuss, by example, comet ablation when the time rate of mass loss is either very small or very large. Little is known about either of these cases, and the following contribution is primarily a series of negative arguments intended to establish the nature and extent of our uncertainties.

2. COMET DUST TAILS

Comet tails are composed of both gas and dust; the relative contribution of each varies greatly from comet to comet. The two components are distinguishable both spectrographically and geometrically. The dust leaves the comet with relatively low velocity and then assumes an orbit different from that of the comet because of solar light pressure. For the smaller dust grains that produce most of the visible dust tail, the effective gravity is 0.1 to 0.5 that on the nucleus itself.

It is a rather simple problem to predict the intensity distribution of a dust tail given a model of the nucleus, the size distribution of the dust, and its optical properties. It is quite remarkable, though, that this problem can also be solved in the opposite direction: Given only the observed tail, one can make a tolerably good estimate of the gas-to-dust ratio, the dust size distribution function, and the rate of loss of material from the comet as a function of time. The procedures for this kind of analysis were developed by Finson and Probstein (1968) and applied to Comet Arend-Roland and a few other comets. The analysis is obviously best applied to the so-called "new" comets that have prominent dust tails. Arend-Roland, for example, was shown to have a dust-to-gas mass-loss ratio of about unity, perhaps an order of magnitude greater than older comets that have made numerous close approaches to the sun. There appear to be only two general possibilities to explain the occurrence of an outer dust shell of new comets: 1) The dust is preferentially accreted during the years since the comet's formation, or 2) ice is preferentially lost.

The first category includes the possibility that the accreted dust constitutes the remains of the solar nebula and represents the final growth of a comet. This ad hoc explanation would be interesting to pursue only if no other mechanism were feasible. Accretion from galactic-dust sources is negligible unless the velocity of the comet, with respect to dust clouds or the general interstellar medium, is high enough for the comet to sweep out large volumes. But if the velocity is high, the dust is destroyed by the collection process. Suppose, though, that the impacting dust serves the primary purpose of selectively removing the volatile ices and thus increasing the concentration of the comet's own dust at the surface. The computation of the maximum effect is trivial, but the required physical parameters can only be estimated. If we take a dust density in the interstellar medium, as given by Allen (1963), of 2×10^{-26} g cm^{-3}, and the solar peculiar velocity, 20 km sec^{-1}, then the energy absorbed by collision in $4 \cdot 10^9$ years is sufficient to vaporize only about 1 mm from a sphere of ice initially at 0°C. The effect falls short of the reality by 3 or 4 orders of magnitude.

The preferential ablation of ices by other means is clearly a possibility. If we arrange the comet temperature to be just sufficient to vaporize, say, 10 m of ice in $4 \cdot 10^9$ years, the gas flow will be insufficient to move even the smallest observable solids in opposition to the gravity of a 1-km body. Thus, there exists a family of solutions to the mechanism of producing dusty comets, limited only by the condition that the vaporization always proceed too slowly to remove dust after the shell is formed and before the comet reaches the inner solar system.

3. THE GEMINID METEOR STREAM

The choice of the Geminids as an example of cometary debris is admittedly some-what peculiar, since this stream is atypical in most respects. First, the Geminid orbit is unlike that of any known comet — aphelia are only about 2.6 a.u., perihelia are 0.14 a.u., and the inclinations (23°) are rather high for a short-period comet. An acceptable mechanism for producing comets of such small aphelia has never been specified. (Perturbations by the planets interior to Jupiter are improbable but perhaps necessary if the nongravitational forces caused by comet outgassing are not sufficient.) Second, the Geminid meteoroids have been shown (Jacchia, Verniani, and Briggs, 1967) to have a higher bulk density than most shower meteoroids. Third, there is less dispersion among the Geminid meteor orbits than among those within most showers, suggesting that the shower formed recently.

And, finally, the Geminids are the only major meteor shower for which there is no associated comet, a fact of particular interest.

All in all, it might seem that one would be more comfortable with the hypothesis that the Geminids are not cometary debris. But the problem of the disruption of a sizable body into rather fine pieces is not solved by giving the reputed parent body a new name.

A crude but instructive estimate of the mass within the Geminid stream can be made from existing observations. The flux of the stream has been determined by radar observations (Hughes, 1973) and by visual techniques (Kohoutek, 1959). Radar records the line density of electrons, a, in the ionized meteor trail. The number of electrons produced is nearly proportional to the mass of the meteoroid, but the proportionality constant is not well known. However, the impersonal radar is well suited for determining the relative distribution of meteor masses. The number of meteors producing an electron line density greater than a is represented by the expression

$$N(a) \propto \frac{c_1 a^{1-s}}{s - 1} \quad . \tag{1}$$

For the Geminids, Hughes gives $s = 1.76$ or, if the mass m is

$$m = K_1 \alpha \quad , \tag{2}$$

then

$$N(m) = \frac{c_2 \, m^{1-s}}{s - 1} \quad . \tag{3}$$

The observed value of c_1 together with the assumed proportionality constant K_1 can give c_2, but here I prefer to use the visual observations to avoid the uncertainty in K_1. The distribution function for meteor magnitudes is usually written as

$$dN(M) = c_3 \, r^M \, dM \quad , \tag{4}$$

where dN is the number of meteors between magnitudes M and M + dM. If the meteor mass is proportional to its intensity I, then

$$m = K_2 \, I = K_2 \, 10^{-0.4M} \tag{5}$$

from the definition of magnitudes. It is readily shown that the magnitude- and ionization-distribution indices r and s are related by

$$s = 1 + 2.5 \log r \quad . \tag{6}$$

I now assume that the radar observations, together with equation (6), properly define the magnitude-distribution law. Of course, this is more likely to be true if the two sets of observations encompass the same mass range, since there is no assurance that the indices s and r are constant for all masses. Hughes estimates that his radar meteors range from 10^{-4} to 10^{-3} g. The estimate was based primarily on the value of K_1 given by Verniani (1966). There is now some indication (Cook et al., 1973) that this was underestimated by perhaps a factor of 10. If so, the mass range of the radar meteors is comparable to that of Kohoutek's visual meteors.

Kohoutek gives the spatial density of meteors with $M_V \leq 3$ as $N(M_V \leq 3) = 2 \times 10^{-8}$ km^{-3}. From equation (6) and from Hughes' value of $S = 1.76$, we find $r \approx 2$.

Integrating equation (4),

$$N(M \leq 3) = \int_3^{-\infty} c_3 \, r^M \, dM = 2.10^{-8} \text{ meteors km}^{-3} \quad , \tag{7}$$

then

$$c_3 = 1.7 \times 10^{-9} \text{ meteors km}^{-3} \quad . \tag{8}$$

If we extrapolate the distribution law to the bright meteors observed by Prairie Network, we find that one camera of that network, which observes about 10^5 km^2 of the meteor region at 100-km altitude, should record about 20 Geminids brighter than -4.0 mag during a 3-hour exposure. In fact, the cameras observed only about one-fourth this number. The correct prediction is made if we change the index r in the magnitude distribution from 2 to 3 for $M_V \leq 0$. This assumption then defines two regimes:

$$N \, dM = \frac{c_3 \, r_A^M}{\ln r_A} \, dM \quad , \quad M_V \geq 0 \text{ (regime A, } r_A = 2) \quad , \tag{9}$$

and

$$N \, dM = \frac{c_3 \, r_B^M}{\ln r_B} \, dM \quad , \quad M_V < 0 \text{ (regime B, } r_B = 3) \quad . \tag{10}$$

The mass density at the maximum of the Geminid shower is then given by the integral of the typical mass of a meteor in the magnitude range dM multiplied by the number at this magnitude. The mass of a meteor can be determined if the integrated photographic intensity is known; i.e., the intensity recorded by a blue-sensitive emulsion. Jacchia et al. (1967) give the mass as 0.02 g for a Geminid whose maximum photographic magnitude, $M_{p \, max}$, is 0.0 mag (meteor no. 9451). In their computations, they assumed the then-current value of the luminous efficiency, which relates mass to luminosity. Two corrections have since been suggested, each of which increases the efficiency by a factor of 2, reducing the mass by the same factor. The first was derived from artificial meteor experiments (Ayer, McCrosky, and Shao, 1970) and the second, from the observation of the Lost City Meteorite (McCrosky et al., 1971).

The photographic intensity differs from the visual intensity (that perceived by the eye), and additional correction factors are necessary. Jacchia (1957) studied this color index (CI) effect and his results can be summarized as follows:

$$CI = M_p - M_v = -1.0 \text{ for } M_p \gtrsim 1 \quad,$$

$$CI = M_p - M_v = -1.8 \text{ for } M_p \lesssim -1 \quad.$$

For simplicity, I take CI = -1.0 for all meteors in regime A and CI = -1.8 for all meteors in regime B. The error so introduced is small compared to other uncertainties to follow. With all the above corrections applied, equation (5) becomes

$$m = K_2(A) \times 10^{-0.4M_v} \qquad \text{(regime A, } K_2 = 0.0125) \quad, \tag{11}$$

$$m = K_2(B) \times 10^{-0.4M_v} \qquad \text{(regime B, } K_2 = 0.026) \quad. \tag{12}$$

Finally, then, the mass density is

$$m = m_A + m_B = \int_0^{M_+} \frac{c_3 K_2(A) r_A^{M_v} 10^{-0.4M_v}}{\ln r_A} \, dM \tag{13}$$

$$+ \int_{M_-}^0 \frac{c_3 K_2(B) r_B^{M_v} 10^{-0.4M_v}}{\ln r_B} \, dM \quad. \tag{13}$$

Because of the choice of indices for the faint and bright meteors, the upper and lower limits do not much affect the value of the mass. If we choose $M_- = -10$ and $M_+ = +8$, the total mass density is about 3×10^{-10} g km^{-3}, of which about 60% is derived from the meteoroids in regime B. The computed mass would be about 15% greater if the limits were chosen to be $M_- \to -\infty$ and $M_+ \to +\infty$.

The total mass of the Geminid stream remains unknown since the earth samples only a chord of a stream of otherwise unknown dimensions. For purposes of the following computation, assume that the stream has circular symmetry and that the earth passes through the center. Radar observations by Šimek (1973) suggest that the Geminid maximum for larger meteoroids lasts for as long as 24 hours and that the rate falls off approximately linearly before and after that time, reaching a near-zero rate 3 days before and after the center of the period of maximum. The earth passes through the Geminid stream at 30 km $\sec^{-1}$ and the heliocentric velocity of the Geminids (nearly perpendicular to the earth's orbit) is 34 km $\sec^{-1}$. The total mass per kilometer along the Geminid orbit at 1 a.u. would then be

$$\text{Mass km}^{-1} = \mathcal{M} \left[\pi r_1^2 + 2\pi \int_{r_1}^{r_2} \frac{(r_2 - r)}{(r_2 - r_1)} \, r \, dr \right] = 2.4 \times 10^4 \text{ g km}^{-1} \quad,$$

(14)

where

$$r_1 = 1.3 \times 10^6 \text{ km}$$

$$r_2 = 6r_1 \quad .$$

The period of the Geminid orbit is about 1.7 years, so the total material in the stream would then be 35 km $\sec^{-1} \times 5.3 \times 10^7$ sec $\times 2.4 \times 10^4$ g km^{-1} = 4.4×10^{13} g.

This computation places no limits on the actual mass in the Geminid stream since, depending on the true cross section of the stream, very much more or very much less mass is possible. Suppose for the moment that the mass is as computed above. There appear to be two possibilities: Either a moderate-sized body — e.g., 440 m in diameter and of density 1 — disrupted entirely, or a larger body shed a considerable shell. If the latter is true, then one wonders why the comet has remained unobserved. A. F. Cook has suggested that an attempt to discover the Geminid parent body, or any larger fragments that remain, is not an insuperable task since the orbit is both small and known with some degree of certainty.

As long as the Geminid parent comet remains undiscovered, there will always be the tantalizing possibility that, somehow, comets are capable of a truly catastrophic disruption into meteoroid-sized bodies.

4. COMET BURSTS

The Geminid meteor shower may or may not be the result of a catastrophic disruption of a comet, but there are numerous other cases of very rapid cometary disintegration. These take two forms. In one, the comet splits into two or more major fragments, and in the other, a sudden increase in the comet brightness occurs, presumably because an outer shell of the comet is expelled. Comet Schwassmann-Wachmann 1 (S-W) is the classic case of this latter behavior. It occasionally shows hundredfold and tenfold increases in intensity. The statistics on these events are poor, but the flares probably occur several times a year. Brightenings by a factor of 2, lasting for a day or so, could occur two or three times more often and escape detection.

During its recent apparition, Comet Tuttle-Giacobini-Kresak (T-G-K) produced a flare that far outdid anything yet observed in Schwassmann-Wachmann. Just before perihelion passage, when the comet was at a solar distance of 1.2 a.u., it increased in brightness by 10 mag in less than a week, perhaps only a day (Marsden, 1973). It became a naked-eye object. There was probably a consensus among comet observers that this event signalled the demise of T-G-K. Indeed, the comet faded rapidly and reached nearly its normal brightness a few weeks after the outburst. But then the comet again produced a flare comparable to the first event. This flare was known to have a rise time of less than three days. C.-Y. Shao and I photographed the comet on July 4.12 with the 155-cm f/5.1 telescope at Harvard's Agassiz Station and found it to be 14.5 ± 0.5 mag. On July 6.87, M. Antal first observed the new outburst. The comet was about 5.0 mag, with a 4-arcmin coma. On July 8.08, J. Bortel observed a 7.5-arcmin coma with binoculars. Shao and I again photographed it on July 8.12 and 9.12; the coma was certainly recorded at diameters of 3 and 5 arcmin, respectively,

and may have been faintly recorded 2 arcmin beyond these limits. The minimum
expansion velocities implied by these observations are given in Table 1.

The velocities are of the order of gas-expansion velocities and are similar to those
reported for other comet outbursts. The coma, except for the central core of about
0.5-arcmin diameter, was predominantly blue. This is characteristic of a gaseous
coma. Spectroscopic observations by Swings and Vreux (IAUC No. 2556) showed CH,
CN, C_2, and C_3 on July 7 and 8. There was only a modest continuum. In this respect,
T-G-K differed from S-W, where the outbursts are predominantly continuum.

Whitney (1955) was apparently the first to point out that the energy contained in
the expanding shell of a S-W outburst might be nearly comparable to the solar flux
falling on the nucleus during the interval between flares. One possible model of the
flaring process, then, would depend on the most volatile compounds vaporizing beneath
the surface of the comet and being contained there by a nearly hermetic outer shell
until the gas pressure caused a catastrophic failure of this shell. In the case of S-W
at some 6.5 a.u. from the sun, any water ice expelled during the outburst would
remain a solid and give rise to the observed continuum radiation of the coma. Similar
particles ejected from T-G-K at 1 a.u. would vaporize far more rapidly. If methane
and ammonia were dissolved in the water ice, these would be released as the parent
molecules of the CH- and CN-band radiation. This model contains a transparent
difficulty. The mass of the driving gas must be comparable with that of the solids if
the solids are to reach the gas-expansion velocity. The volatile material that caused
the outburst must then be "optically inert," if the same substance is responsible for
both S-W and T-G-K outbursts. One possibility for a common, volatile substance that
is not an important radiator in the visible is H_2, but it seems prudent to search for
other possibilities as well.

If the two comets are similar, either methane or ammonia can be the volatile
that causes the outburst (presuming these compounds to be the parent molecules of CH
and CN) if these molecules dissociate in a time scale of days at 1 a.u. but not at 6.5 a.u.
This is true for methane. Strobel (1969) tabulates the absorption cross section for

methane and the ultraviolet solar flux at Jupiter as a function of wavelength. By summing his table, I find a lifetime against photodissociation of about 40 days for CH_4 at Jupiter's distance. The same molecule then has a lifetime of 60 days at the distance of S-W and only 2 days at 1.2 a.u., the distance of the T-G-K outbursts. Methane is not excluded as the driving gas for comet outbursts. Note also that the formation of the continuum by ice grains rather than by dust particles remains a possibility. The higher albedo of ice will alleviate the energy-balance problem of Whitney's dust model.

5. COMET SPLITTING

If comet outbursts are caused by internal pressures, then perhaps comet splitting should be considered as an extreme form of the outburst phenomenom. The only evidence to support this is negative, in that no other model seems capable of explaining all the splitting phenomena. Splitting usually occurs at small solar distances ($r \leq 0.1$ a.u.) and both tidal disruption and thermal shock have been proposed as mechanisms (see, e.g., Whipple and Stefanik, 1965). However, at $r = 0.1$ a.u., the tidal forces are insufficient. The prospects for thermal stress have not been investigated, and I offer the following idealized analysis to provide some insight — though not an exact solution — to this very complex problem. Timoshenko and Goodier (1951) give a summary of time-dependent thermal stresses experienced by an isotropic, isothermal sphere suddenly subjected to external heating. As the comet approaches the sun, particularly from great distances, it may be nearly isothermal, but no comet model would claim that the body was isotropic in a mechanical sense. But the probable heterogeneous nature of a comet nucleus cuts both ways in a discussion of thermal shock. On the one hand, one would expect that the body would fail locally at regions of weakness. On the other hand, such failures would decrease the stress on the interior parts of the body, since it is the homogeneous expansion of the entire body that produces the stresses nearer the center. As an example, it can be shown that a 10-cm spherical rock plunged into boiling water will eventually suffer a central thermal stress in excess of the tensile strength of the material; however, experience shows that a rock so treated will not necessarily break. I presume that some noncatastrophic strain relief takes place in that case, that similar effects may be important in comets, and that their presumed heterogeneous nature protects them against thermal stress rather than enhancing the possibility of failure.

From Timoshenko and Goodier, we find that the maximum tensile stress occurs at the center of a body after a time

$$t = 0.0574 \frac{R^2 c \rho}{K} \quad .$$

For a comet of water ice, I take

$R = 5 \times 10^4$ cm (comet radius)

$c = 0.5$ cal g^{-1} $°C^{-1}$ (specific heat)

$\rho = 0.92$ g cm^{-3} (density)

$K = 5 \times 10^{-3}$ cal sec^{-1} cm^{-1} $°C^{-1}$ (thermal conductivity)

and find

$$t = 1.3 \times 10^{10} \text{ sec} \approx 400 \text{ years} \quad .$$

Thus, no comet will experience a sudden maximum central thermal stress, since it will approach equilibrium at aphelion.

The magnitude of the maximum tension is independent of the body radius and is given by

$$\sigma = 0.77 \frac{\alpha E}{2(1 - \nu)} (T_1 - T_0) \quad , \tag{15}$$

where

$\alpha = 5.1 \times 10^{-5}$ $°C^{-1}$ (coefficient of thermal expansion)

$E = 10$ dynes cm^{-2} (?) (Young's modulus)

$\nu = 0.35$ (Poisson's ratio)

$T_0 = ?$ (initial body temperature)

$T_1 = 0°C$ (final surface temperature) .

I have not found either Young's modulus or the tensile strength of ice, (even though Poisson's ratio is known) but, if the reasonable estimate for E given above is accepted, then the maximum tensile stress will be 2.6×10^8 dynes cm^{-2}, if $T_0 = -100°C$. This is about 1 order of magnitude greater than the tensile strength of rock and perhaps 2 or 3 orders greater than ice. (I recall that Whipple once established, on a napkin at lunch, a lower limit on the tensile strength of ice by noting that icicles are self supporting under gravity g. For a conical icicle of arbitrary angle of density ρ and of length $L = 3$ m, the tensile strength exceeds $g\rho L/3 = 10^5$ dynes cm^{-2}.) We need, then, a better estimate of the actual tensile stress experienced by a comet during perihelion passage.

The radial tensile stress in an isotropically heated body is given as

$$\sigma_r = \frac{2\alpha E}{1-v} \text{ (mean T of entire sphere - mean T inside sphere of radius r)} \quad .$$

$$(16)$$

If we assume that all the solar flux is used to heat an outer shell and that the sphere interior to this shell remains at the initial temperature T_0, then we imply an extreme radial stress. Assume also that vaporization of water ice controls the temperature of the shell at 0°C. (Actually, vaporization keeps the surface somewhat below freezing.)

The term in parentheses in equation (16) is then

$$\frac{r^3 T_0 + (R^3 - r^3) T_1}{R^3} - T_0 = \Delta T \left(1 - \frac{r^3}{R^3}\right) \doteq \frac{3\Delta T \delta}{R} \quad \text{for} \quad \delta \ll R \quad , \qquad (17)$$

where

$\quad$ r $\quad$ = radius of unheated interior at $T = T_0$

$\quad$ δ $\quad$ = R - r, the shell thickness

$\quad$ T_1 = temperature of heated shell

$\quad$ $\Delta T = T_1 - T_0$ $\quad$.

For $\sigma_r = 10^6$ dynes cm^{-2}, the other constants as given previously, and T = 100°C, the heat pulse need penetrate only about 30 cm for the entire inner region of a comet of radius 1 km to be stressed to its nominal failure point. But it is important to note that the strain is also relieved everywhere whenever the failure occurs. Presumably, this happens near the outer limit of the sphere where the material is least compressed initially by gravity.

The idealized analysis suggests that thermal shock is only a skin effect for comets if they are weak structures: Only if they are strong can tensile-limit stresses be generated deep in the body.

6. REMARKS

If it could be demonstrated that the outer shell of dusty new comets was the result of the final accretion process, we would have a new tidbit of information on the condition in the solar system at that time. But it is likely that dust shells are produced simply by the slow sublimation of surface ices. The condition necessary for this mechanism can be anything from a short heat pulse of initial radioactive decay to a steady sublimation at constant temperature over several eons.

The more violent forms of comet ablation — splitting and ejection of shells — would seem to entail mechanisms of greater significance to the eventual understanding of the nature of cometary nuclei, but the mechanisms are still unspecified. I am nearly persuaded that thermal shock is not a factor in comet splitting. The rapid degradation of large areas of some comets' surfaces could result from either thermal shock or internal gas pressure. The two mechanisms are competitive, in that the latter requires a hermetic seal of exterior ices that the former will cause to rupture.

The final answer to this problem must take into account the fact that shell expulsion is limited to a few comets and is often repetitive.

REFERENCES

Allen, C. W., 1963. In Astrophysical Quantities, 2nd edition (Athlone Press, London) p. 252.

Ayer, W. C., McCrosky, R. E., and Shao, C.-Y., 1970. Smithsonian Astrophys. Obs. Spec. Rep. No. 317, 40 pp.

Cook, A. F., McCrosky, R. E., Southworth, R. B., Williams, J. T., Shao, C.-Y., 1973. NASA SP-319, p. 23.

Finson, M. L., and Probstein, R. F., 1968. Astrophys. Journ. 154, 353.

Hughes, D. W., 1973. Mon. Not. Roy. Astron. Soc. 161, 113.

Jacchia, L. G., 1957. Astrophys. Journ. 62, 358.

Jacchia, L. G., Verniani, F. F., and Briggs, R. E., 1967. Smithsonian Contr. Astrophys. 1, 1.

Kohoutek, L., 1959. Bull. Astron. Czech. 10, 55.

Marsden, B., International Astronomical Union Circular No. 2541, June 6, 1973. (Also, IAUCs No. 2542, 2543, 2556, 2559, 2561, and 2568 report on T-G-K outbursts.)

McCrosky, R. E., Posen, A., Schwartz, G., and Shao, C.-Y., 1971. Journ. Geophys. Res. 76, 4090.

Simek, M., 1973. Bull. Astron. Czech. 24, 213.

Strobel, D. F., 1969. Journ. Atmos. Sci. 26, 906.

Timoshenko, S., and Goodier, J. N., 1951. Theory of Elasticity, 2nd edition (McGraw-Hill, New York), p. 416.

Verniani, F., 1966. Journ. Geophys. Res. 71, 2749.

Whipple, F. L., and Stefanik, R. P., 1965. Mem. Soc. Roy. Sci. Liège, ser. 5, 12, 33.

Whitney, C., 1955. Astrophys. Journ. 122, 190.

TABLE 1.

Expansion velocities of T-G-K coma during July 1973 outburst

Time	Observation	Minimum average expansion velocities, assuming $T_0 = 4.12$ (m sec^{-1})
July 4.12	Shao and McCrosky (IAUC No. 2556)	–
July 6.87	Antal (IAUC No. 2556)	330
July 8.08	Bortel (IAUC No. 2559	430
July 8.12	Shao and McCrosky (IAUC No. 2561)	170–280
July 9.12	Shao and McCrosky (IAUC No. 2561)	230–320

DUST IN THE SOLAR SYSTEM

Peter M. Millman

National Research Council of Canada, Ottawa, Ontario

ABSTRACT

Current knowledge of the particulate material of interplanetary space is reviewed in the mass range from 10^{-16} to 10^6 g. Where N is the cumulative number of particles counted down to mass limit m, the slope of the log N vs. log m curve varies between -1.35 and -0.3. For particles less than 1 kg in mass, total flux at the earth—moon system is near 10^{-12} g m^{-2} s^{-1} (2π ster) and rises to about twice this value in the zone between 2.5 and 3.0 a.u. from the sun in the ecliptic plane. The total mass of the interplanetary cloud lies between 10^{19} and 10^{20} g, and over half this mass consists of particles in the mass range $10^{-6.5}$ to $10^{-3.5}$ g, with a peak at 10^{-5} g. The bulk density of the larger particles (>1 g) is low compared to that of the smaller ones (<10^{-6} g). Most of the larger particles are easily fragmented, originate in comets, and are ground to smaller sizes by collisional processes. The mean chemical composition may correspond roughly to that of carbonaceous chondrites, relatively undifferentiated material compared to that of the earth and the moon. Chief areas of uncertainty are in data interpretation for masses >10^4 g and in the mass range from 10^{-16} to 10^{-8} g.

1. INTRODUCTION

In this paper, the term dust will be used in a broad sense to include solid particles over a wide size range, from those with diameters only a few hundredths of a micron up to objects some meters across. In a discussion of dust in the solar system, certain basic questions immediately come to mind: What are the masses and quantities of these particles? What are the relative numbers of various masses, and how are they distributed in relation to the planetary orbits and to the sun? What is the physical structure of the dust and its chemical composition? When, where, and how did it originate?

A few decades ago, most of these questions could be answered only vaguely, as the observational data were more qualitative than quantitative. Recently, however, thanks primarily to the impetus of the space program and its supporting research, our knowledge in this field has expanded rapidly. In fact, new information is arriving at such a rate that any current review is fated to become pretty well out of date in a few years' time. In spite of this, I will attempt to survey briefly some of the present thinking about the particulate matter in interplanetary space.

2. MASS RANGE AND OBSERVATIONAL DATA

Our information about the dust comes from a wide range of observational techniques. Some of the more important of these have been indicated in Figure 1, where the mass range normally related to each type of program is plotted. Too often in the past, data have been quoted without sufficient attention to the particle mass involved. Since, in most cases, shapes and densities are not accurately known, the approximate diameters corresponding to masses are listed on the basis of spherical particles with a standard density of 1 g cm^{-3}. For densities of 10 or 0.1, take the diameters one step up or down, respectively.

Before the advent of the space program, the only information on the smallest particles came from observations of the zodiacal light (van de Hulst, 1947; Leinert, 1971). It was clear from these data that this fine dust was concentrated toward the sun and toward the ecliptic plane, but the observable quantities were not sufficient for the unambiguous determination of the many physical and chemical parameters of the material. Now we have four or five additional techniques that produce data on those particles less than one-millionth of a gram in mass. Both the penetration and the impact of interplanetary dust have been recorded by a large variety of experimental systems mounted on rockets, orbiting earth satellites, and far-roving spacecraft (McDonnell, 1971). The most recent example in the last-named class is Pioneer 10, scheduled to leave the solar system after making close studies of Jupiter and its surroundings (Hall, 1974). The three particle-detection experiments on this vehicle record down to very approximate mass limits of 10^{-12}, 10^{-9}, and 10^{-6} g, respectively. Physical and chemical data result from the actual collection of dust particles, chiefly through the use of balloons and upper air rockets (Hemenway, Hallgren, and Schmalberger, 1972; Bigg, Kvis, and Thompson, 1971). A new and very powerful technique consists of the detailed study of the microcraters found on lunar rock surfaces (Hörz et al., 1974). These data are supplemented by chemical studies of the trace elements in lunar soils and rocks (Anders et al., 1973).

Moving to a consideration of particles with masses appreciably greater than 10^{-6} g, we find that meteor observation is the prime source of information, with a range covering some 14 or 15 orders of magnitude in mass. Among recent summaries of statistical data from meteors are those by Dohnanyi (1972) and Millman (1973). These include observations by radar methods for the smaller particles and by photographic and visual techniques for the larger meteoroids. We are beginning to get statistical data on large meteoroids up to 10^6 g in weight from the passive seismic experiments operating on the lunar surface (Latham et al., 1973). These results are of particular value since reliable data from other sources are very scarce in this size range. Finally, lunar-crater counts for craters whose diameters range from about 100 m up to nearly 10 km (Gault, 1970; Greeley and Gault, 1970) give data on impacting masses greater than 10^6 g — near the upper limit of mass for particles considered in this paper. It is difficult to apply data for craters below 100 m in diameter to a study of the meteoritic complex, as most extended lunar surfaces have reached an equilibrium condition for craters between 0.01 and 100 m and so give little statistical information about the impacting objects.

3. MASS DISTRIBUTION

An important parameter of the dust complex is the relative numbers of particles for various masses. This characteristic is usually described by assuming that over a short range of mass, $N \propto m^{-S}$, where N is the number of particles counted down to a lower mass limit m, and S is defined as the integrated mass index or the negative slope between the plot of log N against log m. Millman (1973) has recently reviewed the observational data that give us values of S for various mass ranges, and the filled-in dots of Figure 2 are taken directly from that paper. More recently, additional studies of microcraters on lunar material have given us interesting information at the low end of the observable mass range (Hörz et al., 1974). At the other extreme of the mass range discussed here, the lunar passive seismic experiment has provided estimates of mass distribution (Latham et al., 1973). These new data are plotted as open circles and broken lines in Figure 2.

The lunar seismic results, which are still preliminary, confirm the lower values for S in the mass range well above 1 g but disagree with the very low values at 10^5 g determined from meteorites and the Prairie Network fireballs. With additional observations from more lunar stations, the statistical value of the seismic data will increase. It is important to remember that results from different techniques may refer to groups of particles with differing physical characteristics, leaving room for a considerable variation of S at any given mass.

When we move down to the mass range from 1 to 10^{-6} g, we find that the sporadic meteors, recorded by a variety of observational methods, give a consistent picture of a steadily decreasing S with decreasing mass, and this is confirmed by the satellite penetrations and the lunar microcraters for masses less than 10^{-6} g. Recent measures of very small microcraters down to those only 0.1 μm in diameter have revealed a second increase in S to values of unity or higher for impacting particles masses below 10^{-10} or 10^{-11} g (Hörz et al., 1974). These have shown up on several of the Apollo 15 rock samples (15017, 15076, 15205) and are plotted at the left of Figure 2. These observations suggest a possible bimodal mass distribution for very small particles, but the origin of this feature is not yet understood.

Preliminary results from Pioneer 10 show low S values beyond distances of 1.5 a.u. from the sun for particles in the 1-g region of mass (Soberman, Neste, and Lichtenfeld, 1974), and these have been plotted as crosses in Figure 2.

4. PARTICLE FLUX

One of the most contentious areas in the study of interplanetary dust, especially in the range of the smallest masses, is that of particle flux. This subject is too complex to give a detailed treatment here, but a brief survey will be attempted so that recent developments can be noted. Traditionally, the unit of flux N is the cumulative number of particles, counted down to a given lower mass limit, that impact on $1 \text{ m}^2 \text{ sec}^{-1}$ from directions integrated over a hemisphere. Flux is normally displayed by plotting log N against log m, where m is the lower mass limit of detection or counting.

The solid lines in Figure 3 are taken from a previous paper (Millman, 1971) and display flux summaries compiled by various authors. Primarily, they cover the mass range from 10 g to 10^{-9} g and are derived from sporadic-meteor observations and satellite penetration data. The point to note is that there is no significant disagreement for the mass range from 1 to 10^{-6} g.

At certain times in the year, the earth passes through more or less concentrated streams of meteoroids where all the objects in each group are moving along a complex of closely similar orbits. These are termed meteor showers, and it has been found that the orbits of certain comets correspond closely to those of given meteor showers, thus indicating a physical connection between comets and shower meteors. At masses of 10^{-1} to 10 g, corresponding to the brighter visual meteors, the meteor showers may have a considerably higher flux than the background of nonshower (sporadic) meteors (Millman, 1970). However, showers generally have a lower S value than the nonshower objects so that the flux from the showers merges into the background flux in meteoroids with masses of 10^{-3} or 10^{-4} g and less.

Examples of recent flux data are indicated with broken lines in Figure 3. The lunar seismic results (Latham et al., 1973) cover the mass range from 10^2 to 10^6 g and thus lie at the right edge of the figure. The preliminary line for the seismic data comes close to the meteor values near log m = 2. For particles some 10^{-6} of the masses recorded by lunar seismographs (that is, meteoroids near mass of 10^{-2} g that correspond to faint meteors near the limit of naked-eye visibility), closed-circuit, television-type electronic equipment has been used for statistical recording (Clifton, 1973). These results agree very well with meteor data from visual, photographic, and radar programs.

Estimated flux curves derived from microcraters on a Luna 16 spherule and rock 12054 (Hartung, Hörz, and Gault, 1972) and on rock 15076.31 (Schneider et al., 1973) have been plotted as examples of some of the better crater counts from lunar material. The effect of the high S values at small particle masses is quite evident for 15076.31. In general, the lunar samples have led to somewhat lower fluxes than those found from near-earth satellite penetrations. It is too early to be sure of the explanation, but among various possibilities are errors in the calibration between crater size and mass, errors in the estimates of time during which a rock surface was exposed, or real differences in the average flux at different periods in the past history of the solar system. It may be significant that flux from lunar Orbiter penetrations is about one-third that from earth-orbiting Explorers.

On the other side of the picture, fluxes calculated from most of the near-earth microphone recordings on satellites and rockets are several orders of magnitude higher than the fluxes found from penetration data (McDonnell, 1971). The discrepancy is particularly marked for particle masses below 10^{-10} g. Various problems exist in eliminating the instrumental noise background from the microphone results, and they will not be discussed further in this survey. The penetration and cratering fluxes seem to rest on a firmer experimental basis. The preliminary values of flux found by Hemenway (1974, this volume), from 36 microcraters on plates exposed on the Skylab S149 experiment, are plotted in Figure 3, and these are reasonably close to the penetration fluxes in both absolute value and variation with mass. As an example of very high fluxes for very small particles, the values found from particle collections made just after the Giacobinid meteor shower in 1972 (Hemenway, 1974, this volume) are also included in Figure 3. One further recent result is a preliminary value found on the HEOS 2 satellite for the cumulative flux in the direction of the apex of the earth's motion with a lower particle mass limit of 10^{-12} g (Hoffmann et al., 1974).

5. TOTAL MASS OF INTERPLANETARY DUST

In a study of the interplanetary dust cloud, it is of interest to find the size range responsible for the greatest percentage of the total mass. Whipple (1967) estimated 40% of the mass in the range of particle masses $10^{-5.5}$ to $10^{-4.5}$ g. Hörz et al. (1974), using lunar microcraters, find the peak mass about an order of magnitude lower near 10^{-6}-g particles. Averaging both mass distributions, and favoring the meteor data above the mass peak and the lunar data below the mass peak, we find that two-thirds of the total mass of the dust complex encountered by the earth is in the form of particles with masses between $10^{-6.5}$ and $10^{-3.5}$ g, or in the 3 orders of magnitude 10^{-6}, 10^{-5}, and 10^{-4} g, respectively.

If we draw a weighted mean flux curve on the basis of the data presented in Figure 3 and integrate the total mass under this curve from 10^{-12} to 10^3 g, we get 6.8×10^{-13} g m^{-2} sec^{-1} over 2π ster, equivalent to 30 tons a day swept up by the earth on the simplified assumption that all areas of the earth's surface receive the same average flux. This absolute value of the total flux is about half that estimated by Whipple (1967) and shows the effect of including the lunar cratering. Anders et al. (1973) give 7.6×10^{-13} g m^{-2} sec^{-1} as the mean influx rate for the three lunar sites of Apollo 11, 12, and 15, a figure based on the estimates of the amount of meteoritic material in the soil. The close agreement with the figure I have given must be mainly fortuitous because, when we see the spread of flux values for small particles exhibited in Figure 3 and we remember that this still does not include the controversial microphone results, we cannot expect to define the overall average flux near the earth–moon system to much better than an order of magnitude. If some of the fine material does not register by penetrating or cratering, it could well be that we should allow for this by raising the flux estimate. It should also be noted that the Prairie Network fireballs (McCrosky, 1968), which are outside the above integration, could add appreciably to the flux estimate if the preliminary mass distribution found by McCrosky is verified by more extensive observation.

The approximate distribution of total mass derived from Figure 3 is given in Table 1. This is similar to that given by Whipple (1967), but with a less prominent peak at 10^{-5} g, giving 55% of the mass in 3 orders of magnitude. The difference arises because the flux plot used here is a continuous curve without the sharp discontinuities present in Whipple's diagram.

It is significant that McCrosky (1958) found, from photometric studies of a meteor wake, that fragmentation of the meteoroid into particles of 10^{-5} g mass would best satisfy the observational data. It would seem that masses near 10^{-5} g are favored in the successive fragmentation of interplanetary material, and this is evident in Figure 2, since it is near 10^{-5} and 10^{-6} g that the integrated mass index S starts to drop steeply down to a value of unity and lower. The bimodal distribution given by the lunar micro-crater data may indicate a second favored area near masses of 10^{-13} or 10^{-14} g, but the fraction of total mass at these very small sizes must be under 10^{-3}. Sekanina and Miller (1973) find that the dust particles in the tail of Comet Bennett peaked near masses of 10^{-12} or 10^{-13} g. The theoretical radiation cutoff near these masses, where the radiation pressure of the sun at 1 a.u. begins to equal the gravitational attraction on the particle, does not prevent the detection of much smaller particles both in lunar cratering and in particle collections. Hawkins (1973) has mentioned the theoretical possibility of a radiation-free component of dust with diameters smaller than the wavelengths of the high optical flux region of the solar spectrum, that is, diameters >0.1 μm. This is an area that holds intriguing challenges for future study, particularly in view of a possible interstellar flux in this regime.

6. FLUX VARIATIONS WITH SOLAR DISTANCE

Very few quantitative data are yet available regarding the variations in the flux of small particles as we move out from the sun. Qualitatively, it had been assumed that the zodiacal-light particles, representing those less than 10^{-6} g in mass, were concentrated toward the sun. Suggested particle space-density laws ranged around the inverse first or second powers of r, the solar distance. It was also realized that there might well be an increase in space density of particles in the asteroid belt, located between r values of 2.0 and 3.6 a.u. Preliminary results from both particle detectors and zodiacal-light recording on Pioneer 10 were presented at the XVI COSPAR meeting in Konstanz, Germany, May 1973. Hanner and Weinberg (1973), using the imaging photopolarimeter to measure the absolute brightness of the zodiacal light between 1.0 and 3.8 a.u. from the sun, find a decrease in zodiacal-light luminosity out to 1.8 a.u. that suggests a particle distribution falling off as r^{-1}. Beyond 1.8 a.u., a slight increase in space density of particles is indicated. These results refer to dust down to mass of about 10^{-12} g.

The optical detection of much larger particles, 10^{-6} to 10^3 g, with the Sisyphus equipment on Pioneer 10, has been summarized by Soberman, Neste, and Lichtenfeld (1974) for solar distances from 2.0 to 3.5 a.u. The data analyzed include 123 recorded events. For particles below 10^{-2}-g mass, the space distribution is fairly uniform. A slight rise in space density, by a factor of about 2, is found for the particles from 10^{-2} to 10^3 g in the area between 2.3 and 3.0 a.u. Since the data for the larger particles consist of only 31 events, the statistical number value is not very large. This increase in the numbers of larger particles relative to the smaller results in the lower S values is plotted in Figure 2. According to Soberman (private communication, 1973), the near-earth value of S had changed appreciably by the time Pioneer 10 had reached 1.2 a.u. from the sun.

The third particle-detection experiment on Pioneer 10 involved the penetration of pressurized stainless steel cells (Kinard et al., 1974), the lower limit of detection being near a mass of 10^{-9} g. On the basis of some 65 penetration events, it is concluded that the particle flux varies smoothly as $r^{-3/4}$ between the earth (1 a.u.) and Jupiter (5 a.u.). No evidence of an increase in flux was found in the asteroid belt from 2 to 3.5 a.u., but a marked gap in penetration events occurred between 1.14 and 1.34 a.u. The most striking data of this experiment gave evidence of an increase in particle flux by over 2 orders of magnitude at the closest approach to Jupiter, 130,000 km above the Jovian cloud tops.

The Harvard-Smithsonian synoptic-year, earth-based observations of faint meteors give us orbital information on over 19,000 meteoroids (Southworth and Sekanina, 1973). This recent study presents new data that revise a preliminary analysis by Southworth

(1967). The new observations refer to meteoroids in the mass range 10^{-6} to 10^{-2} g, with a mean mass near 10^{-4} g. The orbital statistics are used to predict the space distribution of particles on a theoretical basis, making allowance for orbits that cannot be observed with ground-based equipment by extrapolating from the data that have been obtained. The conclusion is that the space density of these particles in the ecliptic plane is at a minimum between 0.7 and 0.8 a.u., then increases steadily as we go out through the earth's orbit to a maximum at 2.6 a.u. of about twice the density found at 1 a.u., followed by a decrease until the space density reaches the near-earth value again at 5 a.u. A similar pattern, but at lower space densities, is found for higher ecliptic latitudes. The smoothed mean space density near the earth is estimated as 4×10^{-22} g cm^{-3}, in agreement with Whipple's (1967) figure.

To summarize, evidence from both spacecraft and ground-based instruments indicates a modest increase (about 2 times) in the space density of small particles as we go from 1 to 3 a.u. Most of this increase seems due to a higher relative number of particles with masses $>10^{-6}$ g in the ecliptic zone between 1.5 and 4 a.u. The collision model developed by Southworth and Sekanina (1973) requires that the greater portion of the particles enter the meteoritic complex in just this zone between 1.5 and 4 a.u., so it is not surprising to find mass distributions here that differ from those at 1 a.u. Any theoretical calculation of the differences in S values we should expect to find must involve consideration of the time constants of the various orbit-perturbation and collisional grinding effects, and the answers are not obvious without detailed calculations.

7. METEOROID DENSITIES

Since unmodified, small interplanetary micrometeoroids have not generally been available for study in the laboratory, densities of the particles themselves must be calculated by indirect methods. Where the dynamical parameters of the air path of a meteor are accurately known from two-station photography or multiple-station radio recording, densities can be found by estimating the surface area from the observed deceleration and combining this area with the particle mass. An appropriate model of the upper atmosphere must be chosen, and the mass of the meteoroid can then be found from a study of the absolute luminosity.

Densities of meteoroids, as determined by Verniani (1967, 1969) from the data of the Harvard-Smithsonian photographic meteor patrol, are plotted in Figure 4. These densities are logarithmic means of groups among 324 precisely reduced photographic meteors with meteoroids in the mass range 0.05 to 5 g. The densities are low, with an overall mean density of 0.3 g cm^{-3}, suggesting porous, friable structures – a view supported by the fact that shower meteoroids do not seem to survive atmospheric passage, except as small dust particles in the micrometer-sized ranges. Among the 220 nonshower meteors in Figure 4, 31 exhibited high densities with a mean at 1.4 g cm^{-3}. These all had orbits with aphelion distances <5.4 a.u. Among the shower meteors, the 20 Geminids are unusual, with a mean density near 1.1 g cm^{-3}, and this shower has an aphelion distance of only 2.6 a.u., the smallest of all the showers listed. The mean density of the 84 remaining shower meteors is 0.25 g cm^{-3}, very similar to the value 0.21 g cm^{-3} for the 189 low-density nonshower meteors. It is evident that higher density meteoroids are more likely to be found among those with small orbits that lie entirely within the orbit of Jupiter.

More recently, Verniani (1973) has made an exhaustive study of the physical parameters of nearly 6000 faint radio meteors with meteoroids in the mass range 10^{-4} to 10^{-2} g and mean mass (from a revised ionizing probability) about 10^{-3} g. Here the density values for individual meteoroids are much less accurate than in the case of the photographic meteors. The overall logarithmic mean density is 0.8 g cm^{-3}, and again a high-density component, about 1.5 g cm^{-3}, is found among the meteoroids with aphelion distances <5.4 a.u.

Indirect evidence for the approximate densities of particles in the size range from 10^{-8} to 10^{-14} g is found in the shape of the microcraters observed on lunar rock surfaces (Hörz et al., 1974). These correspond to laboratory microcraters made by hypervelocity impacting particles in the density range of 2 to 4 g cm^{-3}. There seems to be good observational evidence that the majority of the dust particles impacting on the moon in the above mass range are of a density neither so low as water nor so high as iron. On the assumption that the impacting velocity on the moon is 20 km sec^{-1}, the average density of the particles can be taken as near 3 g cm^{-3}.

For still smaller particles with masses 10^{-15} g and less, we have very little information. A surprising number of heavy elements are listed by Hemenway, Hallgren, and Schmalberger (1972) in very small particles collected by means of upper air rockets. If these form any appreciable percentage of the particle mass, densities considerably higher than 4 g cm^{-3} are indicated.

A summary of estimates for mean densities of interplanetary dust encountered by the earth is listed in Table 2. Since all the densities given in Table 2 are found indirectly, it is quite possible that, in the future, some adjustment will be made in the absolute values. However, the trend from lower to higher mean densities as we move from the larger to the smaller particles seems clear. This is a logical sequence if porous, friable material (cometary debris) is continually broken up into smaller and smaller units by various mechanisms. The continued existence of the denser portions of the original material will be favored in a number of ways, thus leading to higher mean densities for the finer dust. There is also evidence that the dust revolving in smaller orbits, which therefore spends more time closer to the sun, has a higher density component that forms an appreciable fraction of the total complex. Disruptive mechanisms such as radiation pressure and solar wind are more intense in the inner part of the solar system and will have the greatest effect on the material with the lowest bulk density.

In this connection, it should be pointed out that there is no reasonable doubt that shower meteors originate in the fragmentation of comets. Since no systematic statistical difference can be found between the average physical and dynamical parameters of shower meteoroids and of nonshower meteoroids in the same mass range, we conclude that the large majority of all meteoroids observed with earth-based equipment is of cometary origin. Even among the meteoroids producing the very spectacular fireballs of the brilliance of the moon and brighter, McCrosky and Ceplecha (1969) have noted that many may have physical properties very different from those of the meteorites recovered on the surface of the earth.

8. PHYSICAL STRUCTURE OF INTERPLANETARY PARTICLES

Apart from the density estimates noted in the last section, we have a little additional evidence regarding the physical makeup of dust particles. As noted earlier, it is difficult to secure an unaltered microparticle of the interplanetary cloud for laboratory study. Most of the particle-collecting programs are carried out in near-earth space, where we have the problems of contamination with terrestrial aerosols. There is no doubt about a steady influx into our atmosphere of interplanetary mass via meteors and meteorites. Increased ionization in the upper atmosphere following a strong meteor shower was observed as early as 1932 (Schafer and Goodall, 1932; Skellett, 1932), while, more recently, space techniques have made possible the identification of specific ions at various heights where meteoroids fragment and vaporize (Narcisi, 1968; Goldberg and Aikin, 1973). Unfortunately, most of the interplanetary material can be modified in form and altered in composition by the atmosphere and by the collecting techniques.

One type of particle collected on upper air balloon and rocket flights consists of intricate chains of small units held together by cohesive forces (Hemenway and Soberman, 1962; Bigg, Kvis, and Thompson, 1971). Structures of this form have generally been referred to as fluffy particles. The component units range in diameter from 0.1 μm to less than one-hundredth this value, while the overall particle may be several micrometers across — up to 10 or 20 μm in extreme cases. Figure 5 illustrates a complex example of a fluffy particle collected by the group in Australia. It has been suggested that these are interplanetary in origin since they are typical of the higher levels in the atmosphere where a terrestrial origin seems unlikely. The case for the fluffy particles still has to be proved, but it is easy to see them as remains of a cometary matrix of silicates and ices.

Other, more compact particles in the micron and submicron range are also collected (Hemenway, Hallgren, and Schmalberger, 1972). In the atmospheric collections, the interplanetary component will consist of the small pieces of much larger meteoroids that have been fragmented and vaporized by aerodynamic forces, plus the true

micrometeoroids (Whipple, 1950, 1951) that are small enough to have their initial geo-
centric velocity reduced before any appreciable ablation has taken place. In contrast,
impacts on the lunar surface take place at roughly the same high velocity (generally
assumed near 20 km sec^{-1}) for both the large and the small meteoroids. The shape
parameters of the microcraters on lunar rocks clearly indicate that the impacting
particles are of a generally rounded, equidimensional form, are not platelets, rods,
nor needles, and have an average length:width ratio of <2 (Hörz et al., 1974).

9. CHEMICAL COMPOSITION OF INTERPLANETARY DUST

The meteoritic flux on the lunar surface quoted earlier, and estimated by Anders
et al. (1973) from a study of key trace elements in the lunar regolith at the Apollo
landing sites, seems to have a composition most like that of the C1 carbonaceous
chondrites, that is, with enriched siderophiles and volatiles compared to the other
components of the lunar soil. The more direct observational evidence for the chemical
composition of particulate material in the solar system has been reviewed in a recent
paper (Millman, 1974) and will be noted only briefly here.

Vaňýsek (1973) has surveyed the observational evidence for the composition of dust
in comets and finds that such fragmentary data as are available favor silicate rather
than metallic material in the submicrometer grains. Both cometary and meteor
spectra show evidence of the presence of light volatiles such as H, N, and O, as well
as C, Fe, and other common elements usually found in meteorites. Unfortunately, in
both groups of spectra, it is difficult to determine quantitative abundances of the elements, but a start has been made at a number of laboratories, and we should have
better data in a few years (Harvey, 1973; Millman, 1972).

Up to the present, nothing seems to contradict the assumption that the great majority of meteoroids with masses greater than 1 g, which we observe as the brighter
meteors, have elemental abundances much like the common chondrites, but with possible enrichment of some of the light elements, as is the case for the carbonaceous
chondrites. In the mass range 1 to 1000 g, nonsilicate meteoroids are certainly in
the minority, less than 3%, and practically unknown among the members of showers
(Millman, 1963). Among 500 faint-meteor spectra with mean mass just over a gram,

Harvey (1973) finds only 1 or 2% of the pure nickel-iron type and 2% of the iron-deficient, magnesium-calcium-rich spectra. In general, meteor spectroscopy gives evidence of a basic uniformity in the chemical composition of cometary meteoroids, peculiar objects being relatively rare. The major differences in observed meteor spectra result from the effects of size and velocity ranges, rather than composition.

10. DISCUSSION

The high S values for mass distribution of the dust in the range 1 to 10^{-6} g seem well determined by a number of completely independent observational techniques. Using a theoretical collisional model of meteoroid fragmentation with constants based on laboratory impact experiments, Dohnanyi (1970) has shown that a particulate complex with $S > 0.8$ will, in general, be unstable unless it is fed by material with an initial S more than twice the above value. Comets, well observed as bodies that fragment and disperse their mass, together with the associated meteor streams, which are also dispersing, would seem to supply the necessary source material to satisfy Dohnanyi's theorem. The hypothesis is strengthened by the friable character of most of the meteoroids encountered by the earth and by the observed mass distribution of the meteor showers that, according to Dohnanyi, should reach a final state where the faint members have an S near 0.5.

Whipple (1967) has estimated the total mass in the dust cloud as 2.5×10^{19} g, and in view of various uncertainties, this is probably as good a value as any to adopt. While it might be lowered a little by lunar surface data, the Pioneer 10 results might raise it. Whipple also estimated the mean lifetime of cloud particles as near 10^5 years and the total input needed to maintain the cloud as between 1 and 2×10^7 g sec^{-1}. This is of the same order of magnitude as that required by Dohnanyi's theory and is quite consistent with the idea that disintegrating comets provide the source. The larger meteoroids, mass >0.1 g, disappear from the meteoritic complex almost entirely as a result of erosion, while the smaller meteoroids, down to masses of 10^{-6} g, are subject to collisional destruction. The Poynting-Robertson effect, a drag on the orbital motion of particles resulting from the reradiation or reflection of solar energy, becomes significant in dust removal of masses $<10^{-6}$ g, where knowledge of the size distribution and other physical parameters is much less certain.

We must not omit mention of a couple of skeletons in the closet. We still lack detailed statistical knowledge of the flux and physical characteristics of masses in the range 10^2 to 10^8 g. If we take the mass represented by McCrosky's (1968) curve for the flux of the Prairie Network fireballs (10^3 to 10^7 g), it is at least half the total mass found for the particles $<10^3$ g. If this mass-distribution curve is extended at all to larger objects, the mass of the bright fireballs is considerably greater than that of all the smaller material. Yet the lunar seismic results give mass totals from 2 to 3 orders of magnitude smaller, more consistent with the fainter meteor data. Both the fireball and the seismic data are preliminary and could be considerably modified by additional observations.

The other area of uncertainty concerns the particles less than 10^{-6} g in mass. The high relative fluxes recorded by microphone and in particle collections have been mentioned already. Hawkins (1973) feels that these data cannot be disregarded completely when analyzing the flux picture, since they have too much internal consistency. Some reasonable answer must be found, but here again data from the lunar surface agree with the lower flux regime, 3 or 4 orders of magnitude below the microphone results.

Thus, at both ends of our flux curve, we find hints that additional mass may be reaching the earth, but it remains unrecorded by various standard observational techniques. The estimates of the cosmic flux on earth, based on a study of deep-sea sediments, have a bearing on the subject. Barker and Anders (1968) have used the measures of iridium and osmium, elements depleted in the earth's crust, to indicate the cosmic component in deep-sea sediments. They find a flux roughly 4×10^{-12} g m^{-2} sec^{-1} and a firm upper limit of about twice this value. This leaves room for from 5 to 10 times the total flux found on the basis of meteor-counting, penetration, and cratering data. It seems rather unlikely that we can raise it by more than an order of magnitude.

REFERENCES

Anders, E., Ganapathy, R., Krähenbühl, U., and Morgan, J. W., 1973. Moon 8, 3.

Barker, J. L., and Anders, E., 1968. Geochim. Cosmochim. Acta 32, 627.

Bigg, E. K., Kviz, Z., and Thompson, W. J., 1971. Tellus 23, 247.

Clifton, K. S., 1973. Journ. Geophys. Res. 78, 6511.

Dohnanyi, J. S., 1970. Journ. Geophys. Res. 75, 3468.

Dohnanyi, J. S., 1972. Icarus 17, 1.

Gault, D. E., 1970. Radio Sci. 5, 273.

Goldberg, R. A., and Aikin, A. C., 1973. Science 180, 294.

Greeley, R., and Gault, D. E., 1970. Moon 2, 10.

Hall, C. F., 1974. Science 183, 301.

Hanner, M. S., and Weinberg, J. L., 1973. XVI COSPAR paper C.3.8, May 1973, Konstanz, Germany.

Hartung, J. B., Hörz, F., and Gault, D. E., 1972. Geochim. Cosmochim. Acta Suppl. 3, 2735.

Harvey, G. A., 1973. In Evolutionary and Physical Properties of Meteoroids, ed. by C. L. Hemenway, P. M. Millman, and A. F. Cook, NASA SP-319, 131.

Hawkins, G. S., 1973. In Space Research XIII, ed. by M. J. Rycroft and S. K. Runcorn (Akademie-Verlag, Berlin) p. 1159.

Hemenway, C. L., Hallgren, D. S., and Schmalberger, D. C., 1972. Nature 238, 256.

Hemenway, C. L., and Soberman, R. K., 1962. Astron. Journ. 67, 256.

Hoffmann, H.-J., Fechtig, H., Grün, E., and Kissel, J., 1974. Planet. Space Sci., in press.

Hörz, F., Brownlee, D. E., Fechtig, H., Hartung, J. B., Morrison, D. A., Neukum, G., Schneider, E., and Vedder, J. F., 1974. Planet. Space Sci., in press.

Kinard, W. H., O'Neal, R. L., Alvarez, J. M., and Humes, D. H., 1974. Science 183, 321.

Latham, G., Ewing, M., Dorman, J., Nakamura, Y., Press, F., Toksöz, N., Sutton, G., Duennebier, F., and Lammlein, D., 1973. Moon 7, 396.

Leinert, C., 1971. In Space Research XI, ed. by K. Ya. Kondratyev, M. J. Rycroft, and C. Sagan (Akademie-Verlag, Berlin) p. 249.

McCrosky, R. E., 1958. Astron. Journ. 63, 97.

McCrosky, R. E., 1968. Smithsonian Astrophys. Obs. Spec. Rep. No. 280, 1.

McCrosky, R. E., and Ceplecha, Z., 1969. In Meteorite Research, ed. by P. M. Millman, (D. Reidel Publ. Co., Dordrecht-Holland) p. 600.

McDonnell, J. A. M., 1971. In Space Research XI, ed. by K. Ya. Kondratyev, M. J. Rycroft, and C. Sagan (Akademie-Verlag, Berlin) p. 415.

Millman, P. M., 1963. Smithsonian Contr. Astrophys. 7, 119.

Millman, P. M., 1970. In Space Research X, ed. by T. M. Donahue, P. A. Smith, and L. Thomas (North-Holland Publ. Co., Amsterdam) p. 260.

Millman, P. M., 1971. Amer. Sci. 59, 700.

Millman, P. M., 1972. Journ. Roy. Astron. Soc. Canada 66, 201.

Millman, P. M., 1973. Moon 8, 228.

Millman, P. M., 1974. Journ. Roy. Astron. Soc. Canada, 68, 13.

Narcisi, R. S., 1968. In Space Research VIII, ed. by A. P. Mitra, L. G. Jacchia, and W. S. Newman (North-Holland Publ. Co., Amsterdam) p. 360.

Schafer, J. P., and Goodall, W. M., 1932. Proc. I.R.E. 20, 1941.

Schneider, E., Storzer, D., Hartung, J. B., Fechtig, H., and Gentner, W., 1973. Geochim. Cosmochim. Acta Suppl. 4, 3277.

Sekanina, Z., and Miller, F. D., 1973. Science 179, 565.

Skellett, A. M., 1932. Proc. I.R.E. 20, 1933.

Soberman, R. K., Neste, S. L., and Lichtenfeld, K., 1974. Science 183, 320.

Southworth, R. B., 1967. In The Zodiacal Light and the Interplanetary Medium, ed. by J. L. Weinberg, NASA SP-150, 179.

Southworth, R. B., and Sekanina, Z., 1973. NASA CR-2316, 106 pp.

van de Hulst, H. C., 1947. Astrophys. Journ., 105, 471.

Vanýsek, V., 1973. In Space Research XIII, ed. by M. J. Rycroft and S. K. Runcorn (Akademie-Verlag, Berlin) p. 1173.

Verniani, F., 1967. Smithsonian Contr. Astrophys. 10, 181.

Verniani, F., 1969. Space Sci. Rev. 10, 230.

Verniani, F., 1973. Journ. Geophys. Res. 78, 8429.

Whipple, F. L., 1950. Proc. Nat. Acad. Sci. U.S. 36, 687.

Whipple, F. L., 1951. Proc. Nat. Acad. Sci. U.S. 37, 19.

Whipple, F. L., 1967. In The Zodiacal Light and the Interplanetary Medium, ed. by J. L. Weinberg, NASA SP-150, 409.

TABLE 1.

Percentage of total mass in various mass categories.

log m	−10.5		−9.5	−8.5	−7.5	−6.5		−5.5		−4.5		−3.5
%		0.1		1	3	8		16	21		18	

log m	−3.5		−2.5	−1.5	−0.5	+0.5		+1.5		+2.5		+3.5
%		14		9	5	3		1	0.5		0.2	

TABLE 2.

Mean densities of interplanetary dust.

Mass range (log m) (g)	Density (g cm^{-3})
-1 to +1	0.3
-4 to -2	0.8
-14 to -8	3.
<-14	>4. ?

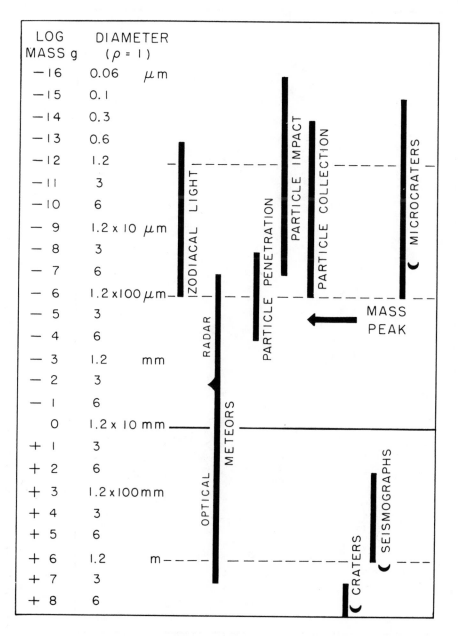

Figure 1. The mass range and size of interplanetary particles observed by various techniques.

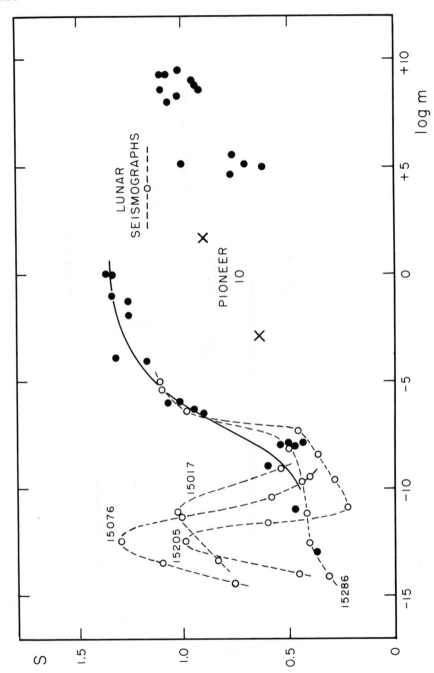

Figure 2.   The integrated mass index S plotted against the corresponding mean log mass in grams. Filled-in dots and solid line are taken from Millman (1973). Numbers at the left refer to individual lunar rocks.

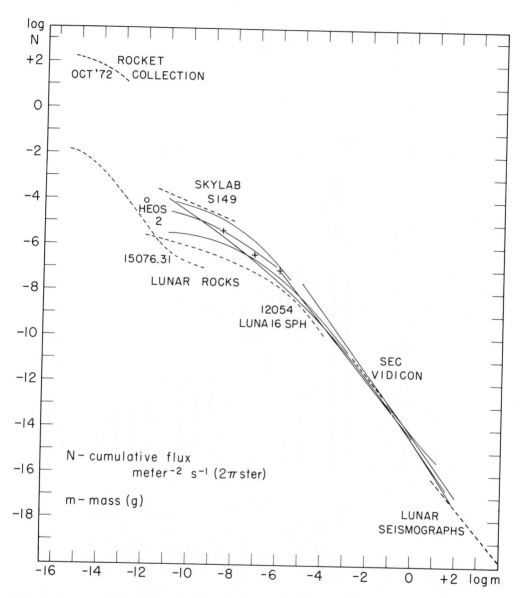

Figure 3. The relation between the cumulative number of particles and the lower limit of mass to which they are counted, as derived from various types of recording. The solid lines are taken from Millman (1971); the crosses are the Pegasus and Explorer penetration data.

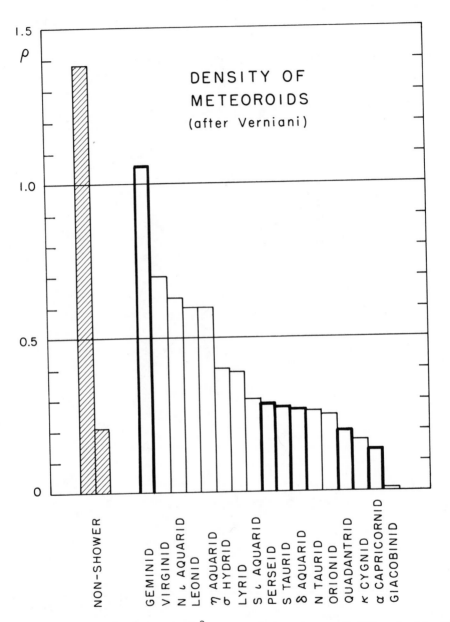

Figure 4. Mean bulk densities (g cm^{-3}) of groups of meteoroids, determined by Verniani
(1967, 1969) from 324 precisely reduced Super-Schmidt photographic meteors.
Shower groups in heavy outline contain seven or more meteors.

Figure 5. Example of a fluffy particle, collected at a height of 27 km over Mildura, Australia, on March 16, 1971. (Courtesy of Z. Kviz and University of Sydney, Australia.)

COLLECTIONS OF COSMIC DUST

Curtis L. Hemenway

Dudley Observatory and The State University of New York at Albany

ABSTRACT

Some recent results of collection experiments in the upper atmosphere by rocket techniques, by balloon collections in the intermediate atmosphere, and by recoverable satellites in near-earth space are described. Evidence is presented for the existence of relatively high and variable fluxes of submicron particles entering the atmosphere. These submicron-particle collections are consistent with implications from zodiacal light studies, provided the particles have a solar origin and are drifting outward with a constant radial velocity. A simple estimate suggests that these particles, by virtue of their refractory nature, can survive ejection from the sun near sunspots.

This paper will discuss some of the results of collection experiments that my colleagues and I have carried out over the past 17 years. It should be noted that the possibility and importance of collecting cosmic dust was suggested to me by Professor Whipple during a sabbatical leave that I spent at Harvard during 1955-1956.

The first task is to provide evidence that small cosmic dust particles are coming into the earth's atmosphere in significant numbers. One way to do this is to use electron microscopy to compare the shape and morphology of the particles collected in some of our rocket collection experiments (Hemenway and Soberman, 1962; Hemenway and Hallgren, 1970; Hallgren et al., 1973), generally in the altitude range of 80 to 140 km, with some of the penetration holes found in ultra-thin films exposed in the altitude range 288 to 320 km during Gemini (Hemenway, Hallgren, and Kerridge, 1968). The reliability of the rocket collection techniques has been enhanced considerably by the development of inflight shadowing (Hemenway and Hallgren, 1970), by which the collected particles are tagged to identify the altitude interval within which they were collected.

Figure 1 shows a particle collected with our Pandora II rocket techniques in 1970 in northern Sweden during a noctilucent cloud (NCL) display. It consists of a low-density, weak structure surrounding a high-density core. The core appears to be made up of high Z elements that are quite opaque to 100-kV electrons by virtue of containing relatively large numbers of electrons per unit volume. Figure 2 shows a penetration hole obtained during Gemini 9 in approximately 400-Å thick nitrocellulose films. This hole appears to have been created by the penetration of the high-density core while the weak surrounding material is left behind on the thin film. These dual-component particles have been found on many rocket collection flights. In some cases, almost all the particles collected are of this dual-component nature (Hemenway, Soberman, and Witt, 1964; Hemenway and Hallgren, 1969). It should be noted that studies of rocket particles with the electron microscopes available to us 10 years ago had significant internal electron scattering, so that the high atomic number and probable high-density nature of the cores of these particles may not have been apparent.

Figure 3 shows a cluster of particles that apparently broke up before being collected during a rocket collection at White Sands, New Mexico, on August 12, 1968. Figure 4 shows a cluster of penetration holes found in a thin nitrocellulose film exposed during Gemini 9. These pictures suggest that some of the particles entering the atmosphere are very fragile and are easily broken up, possibly by chemical interaction with atomic oxygen in the ionosphere.

Figure 5 shows a 1-μ particle collected by Pandora II surrounded by about 20 particles in the 50- to 100-Å size range. Figure 6 shows a penetration hole structure surrounded by about 200 small penetration holes collected during Gemini 9. We note that so far, no individual particles have been seen that are less than about 600 Å in diameter. Instead, the ultra-small particles always appear to be a part of a larger structure.

Additional support for the concept of small particles entering the earth's atmosphere is found from the fact that the fluxes of particles collected as a function of altitude appear to be approximately constant as shown by the example given in Table 1. The fluxes are obtained by dividing the number of particles per size interval per unit area by the effective collection time, i.e., the time computed for them to fall through the altitude interval of the collection (Soberman and Hemenway, 1965). The computer program for the effective collection time involves three domains: free molecular flow at high altitude, Epstein's formula when the particle has slowed down, and Stoke's Law at lower altitude. In 3 out of the 25 rocket flights, evidence of significant layering has been found. It may be of interest that no evidence of particle concentration as a function of altitude has been found from noctilucent cloud collections thus far. Furthermore, cooperative rocket particle collection and impact-velocity measurements have found all particles detected to be falling downward generally at about the computed falling velocities (Rauser and Fechtig, 1972). Thus, for several independent reasons we are confident that we can reliably detect and study particulate materials entering the earth's atmosphere from beyond.

We next show some additional particles collected during NLC displays in Sweden. Figure 7 shows a dual-component particle with a high-density core that appears to have been partially melted. The 0.1-μ high-density particle shown in Figure 8 appears to have also experienced a severe thermal environment, in which it has been partially melted and flared back. We find evidence in the study of submicron particles collected in the outer fringes of our atmosphere that many of the particles have experienced a more violent heating than would be expected from the heating upon entry into the earth's atmosphere, as described by the Whipple micrometeorite theory (Whipple, 1950).

In Figure 9, a shadowed, dual-component particle is shown from our first noctilucent cloud collection (Hemenway, Soberman, and Witt, 1964); this particle has been affectionately called "Foo Foo" by cosmic-dust workers. The eyes, nose, and mouth are formed by parts of the high-density core, the hat is formed by the core shadow, and the mottled structure surrounding the core appears to represent a chemical interaction of the fragile, low-density mantle material with the upper surface of the nitrocellulose collection film during the 1-month interval between the collection and the laboratory shadowing.

We have tried many times to collect enhanced numbers of submicron particles by rocket techniques at times of meteor-shower activity or at times of the earth's transit through the plane of a comet's orbit and, in all cases but one, have failed to find significant enhancements. An example of such a failure is shown in Figure 10 where a particle is shown that is representative of a large number of particles collected on November 18, 1965, at White Sands, New Mexico, by our Pandora I system (Hemenway and Hallgren, 1969). Initially, we thought these particles were from the Leonid Meteor Stream; however, subsequent computations and work have indicated that these core-mantle particles probably had little to do with the Leonids and that they are similar in nature to the particles shown in Figures 1 and 7, except that the micrographs were taken with older electron microscopes.

The one circumstance in which we appear to have collected material during the transit of the earth through a comet's orbital plane was during the hoped-for Giacobini Zinner event (GZ) on October 8, 1972. Figure 11 shows the integral size distributions of particles collected per unit area on that date (at about the time of the GZ node), on October 9, and also on October 12. Since the altitude intervals of the collections were closely matched, it appears that a significant enhancement of particles was collected at the time of the GZ node. Figures 12 and 13 show particles collected on October 8.

Figure 12 shows a typical loose conglomerate particle of a type to be expected as residue from the Whipple dirty-ice comet model. The particle in Figure 13 is roughly spherical and is more transparent in the center than near the edge. This latter particle may be the residue of an icy particle in which the impurity constituents were concentrated at the particle's roughly spherical surface by evaporation of a volatile component, thus resulting a spherical shell distribution of the residual material. The particles shown in Figures 12 and 13 are not of a high-density nature, although a few high-density, core-mantle particles were found. The studies of these materials are continuing, and this is the first publication of our Giacobini Zinner node collections.

To collect larger cosmic-dust particles (i. e., $> 5 \mu$), balloon techniques are preferred since the area-time products of the collections are generally much greater than those for rocket collections. We have been using a settling-plate technique mounted on top of zero-pressure balloons within a sealable box. This technique, which we have named Sesame, has been flown on 34 balloon flights (Hemenway, Hallgren, and Coon, 1967; Hemenway et al., 1971).

Figure 14 shows a scanning microscope picture of a particle collected during our Sesame XXVIII (August 13, 1968) balloon collection flight. It is about 15μ in diameter and is probably of a stony nature. Most large particles collected show no evidence of surface melting; however, about 5% do. Figure 15 shows a particle collected during Sesame XXV (November 18, 1967) in which surface melting appears to have played a detectable role, and Figure 16 shows a particle from Sesame XXXIV (August 13, 1970) in which the particle ablation appears to have been more intense. Thus, a few sufficiently large particles do show some evidence of surface ablation, as expected by the Whipple micrometeorite-entry theory. In the intermediate sizes, a negligible number of particles show evidence of melting whereas, in smaller submicron sizes, a significant number of particles show evidence of partial melting.

We have attempted to find evidence of crystal structure within the Sesame particles by using a sensitive x-ray diffraction technique and found that 13 out of 14 particles $> 20 \mu$ showed no detectable evidence of internal crystal structure, whereas smaller control polycrystalline and single-crystal particles consistently did. This appears to suggest, as did our earlier electron-diffraction studies of the Venus Fly Trap (VFT) particles (Hemenway and Soberman, 1962), that crystal structures are rare within the cosmic-dust particles studied thus far. Both the large particles collected during the Sesame program and the VFT particles appear to be the result of atom-by-atom accretion under cool conditions. More work should be done concerning crystal structure of cosmic-dust particles.

The Sesame and VFT particles were sufficiently large that composition studies using standard electron-beam-probe techniques could be employed. Silicon was the most common element observed in the Sesame particles, with carbon, oxygen, aluminum, magnesium, calcium, and iron frequently found (Hemenway et al., 1971). The major anomaly in the VFT collection was that titanium appeared to be more abundant than expected (Hemenway and Soberman, 1962).

Figure 17 shows portions of two x-ray spectra taken with our Phillips electron microscope used as an electron-beam probe. The upper spectrum was obtained from the small high-Z particle shown, and the lower spectrum was taken about 30 µ away from the particle. We see that a weak x-ray line, probably from hafnium, has appeared in the spectrum of the particle. In similar studies of 42 high-density particles from our noctilucent cloud collections of 1970 and 1971, weak x-ray peaks were observed that appear to be associated with such elements as hafnium, tungsten, lanthanum, and tantalum. On the other hand, silicon was observed only twice in these 42 particles.

Since the one common characteristic of the high-Z elements found to be abnormally abundant in the high-density cored particles is their high temperature stability, and since we have been unable to find an alternate source for them, we have concluded that most of the dual-component particles we have been observing have a solar origin (Hemenway, Hallgren, and Schmalberger, 1972). This concept is consistent with Roach et al.'s (1954) observations, which show that the zodiacal light brightness as a function of elongation angle is continuously connected to the Fraunhofer F corona. Also, Berg's "solar noise" pulses from Pioneer 8 and 9 (Berg and Gerloff, 1971) can be interpreted as detecting outward-flowing particles from the sun. Also, Dubin (1973) has found an explanation of the blue clearings of Mars at times of opposition – that the earth's magnetosphere casts a shadow in the outward-flowing flux of submicron particles.

We next turn to our unpublished, most recent space exposure experiment, wherein cosmic dust particles are detected by their impacts on highly polished metal plates and by their penetration holes in thin films. Figure 18 shows the Skylab S-149 Micrometeoroid Impact Detector as deployed for exposure with blank slides in position. Figure 19 suggests the variety of experiments carried out in each of the four sets of four cassettes to be exposed at a given time. Figure 20 shows the orientations of the collector during the 34-day anti-solar (-z) and the 46-day solar-facing exposures obtained during 1973.

Figure 21 shows a scanning electron micrograph of a cosmic-dust impact crater found on a polished copper slide. This crater is similar to two of the three impact craters found in the S-10 experiment during Gemini. Thus far, 36 impact craters have been found during the first two Skylab exposures by the Dudley Staff, with additional craters informally reported by Dr. H. Fechtig and Mr. E. Fullam, each of whom provided experiments for exposure during the S-149 experiment.

Figure 22 shows a small unusual impact crater found in a electron microscope copper-grid wire covered by a thin film of gold 500 to 800 Å thick. The gold film was 1 to 2 µ above the copper wire. The particle penetrated the gold foil, made a 0.6-µ diameter impact crater in the copper wire, and the material blown back from the crater subsequently made a larger blowout hole in the gold foil.

A number of thin-film penetration experiments were successfully carried out during the S-149 exposures of Skylab. Figure 23 shows an electron microscope penetration hole in 500- to 800-Å gold foil. Figure 24 shows an area of the 500- to 800-Å thick gold foil directly beneath the penetration hole shown in Figure 23. It can be seen that the particle broke up into many smaller pieces on penetrating the top foil. This once again shows the fragility of some of the impacting particles and is a miniaturized working demonstration of Whipple's "meteoroid shield" for the protection of spacecraft.

Figure 25 shows the darkening and flaking of a polished silver slide exposed on the solar side, compared with the absence of an interaction in a stainless steel slide from the same solar exposure. The black material coating the silver slides was found, by x-ray analysis, to be silver oxide. It appears possible that the silver reacted with the atomic oxygen at the 430-km altitude of the exposure.

Figure 26 shows a penetration hole observed in a thin nitrocellulose film (~400 Å) that was given a solar-facing exposure. We believe the direction of the penetrating dual-component particle can be inferred from the direction of the roughly cylindrical central hole. This event is somewhat similar to that shown in Figure 2, except that chemical interaction of the residual material with the nitrocellulose has been minimized.

It is anticipated that the full scientific exploration of the S-149 experiment will take many years and will contribute reliable information on near-earth particle fluxes — size distribution and chemistry, as well as the average directional characteristics of the cosmic-dust particles in the solar system at 1 a.u.

Figure 27 shows an intercomparison of the fluxes of the particles collected in the earth's upper atmosphere with the near-earth space exposures. The preliminary crater fluxes were obtained from the S-149 crater diameters by assuming that the crater diameters are three times the particle diameters; they agree fairly well with the near-earth penetration measurements and the S-10 crater data of Hemenway, Hallgren, and Kerridge (1968). The uncertainties concerning these integrated individual particle fluxes in the mass range from about 10^{-12} to 10^{-7} g are not great, certainly less than 1 order of magnitude. We should note that the anti-solar fluxes of S-149 appear to be significantly greater than the solar-facing exposures, possibly suggesting that these large cosmic-dust particles are spiraling inward toward the sun and that more are intercepted on the anti-solar side. Straight line A represents our preliminary best estimate of the near-earth particle fluxes from studies of craters found on our satellite experiments. A lower limit implied by our S-149 gold-foil penetration studies is also shown in Figure 27.

The flux measurements from recent balloon Sesame collections (Hemenway et al., 1971) continue to show larger fluxes by about 2 orders of magnitude in the mass range from about 10^{-10} to 10^{-8} g. These fluxes are compared with our earlier Sesame control flights (Hemenway, Hallgren, and Coon, 1967), represented by line B. We have tried doing Sesame collections while the balloon is rising from 60,000 to 110,000 feet and have found consistently lower fluxes, which we now believe to be the result of lower collection efficiency; the Sesame sweep-up fluxes approach the Sesame floating fluxes as larger particle sizes are reached.

Balloon collections made one or two days after strong meteor showers such as the Perseids and the Geminids often show enhancement of particle fluxes (Hemenway, Hallgren, and Coon, 1967). The same flux in the mass range from 10^{-10} to 10^{-8} g falling into the floating settling-plate collector is about 3×10^{-11} g m^{-2} sec^{-1}. A typical estimate of the total mass influx from meteoroids would be about 10^{-13} g cm^{-2} sec^{-1}. In view of the uncertainties concerning meteor physics and of our ability to measure accurately the masses of individual particles (Patashnick and Hemenway, 1969), it may turn out that the balloon collections will provide a useful measurement of the total mass flux of crumbled fragments from larger meteor bodies. We find it difficult to believe, despite the absence of inflight shadowing in balloon collections, that our judgement of cosmic-dust particles is wrong most of the time.

Whereas balloon fluxes usually do not vary by more than an order of magnitude, rocket collection fluxes frequently are enhanced by 2 orders of magnitude and have been observed to increase by nearly 4 orders of magnitude. The rocket collections can be made in a much lower mass range than can be measured by other techniques. The recent GZ data are shown along with data from several recent collections and an old noctilucent cloud collection. As noted earlier, the rocket fluxes shown in Figure 27 depend on a computed effective collection time and an assumed density (3 g cm^{-3}). We do not believe these fluxes are individually in error by as much as an order of magnitude. Instead, there appears to be a significant time variability of the influx of submicron cosmic-dust particles (Hemenway and Hallgren, 1970).

The higher magnitudes of the submicron fluxes may represent, in part, fragmentation from larger bodies, dependence of the capture cross section of the earth on the earth's magnetospheric cross section, and a coulomb drag sweep-up phenomenon by the solar wind plasma. The reasons for the enhanced submicron flux are not understood, and we are currently attempting a numerical model for the earth's capture of charged submicron particles, both individually and coupled to plasma clouds. It is interesting that the slopes of the GZ control collection of November 11, 1972, and the NLC collection of August 11, 1972, are similar. Also, two separate flights, through NLC displays on successive days in 1971, gave about the same fluxes. Figure 27 also shows data taken on the same rocket flight (NLC, July 31, 1971) by Rauser and Fechtig (1973) with their plasma detector, as well as that taken by Lindblad (Lindblad, Arinder, and Wiesel, 1973) with his microphone technique. It appears that the collection, plasma detector, and microphone data are telling a self-consistent story. The Lindblad data have been adjusted to allow for the low impact velocities measured by Rauser. It is noted that the high fluxes of submicron particles shown in Figure 27 appear to be necessary to explain the twilight photometric measurements of Link (1973). It is also interesting that the typical minimum values of rocket collection fluxes, when converted to particle numerical densities, predict the right order of magnitude for the numerical densities of submicron particles in the stratosphere, and also, by virtue of the variability of the particle-flux input, predict the "layering" observed at stratospheric altitudes (Rosen, 1969).

The variability of the influx of cosmic dust appears to provide at least a partial explanation for noctilucent clouds. The increased numbers of incoming particles serve as nucleating centers on which the supercooled water vapor deposits at the temperature minimum of the mesopause. The particles grow to such a size that the total scattering cross section at the 84-km altitude layer is greatly increased and becomes visible long after sunset and long before sunrise, while the 84-km region is still illuminated by the sun. Thus, it is not surprising that we do not find from collection studies a layer of particle concentration at the noctilucent-cloud level. Instead, there is evidence suggesting that, if anything, we see a deficiency of particle numerical density that results from the fact that the larger ice-laden particles fall faster. The transparent "water-reaction spots" observed in protected calcium films in 1962 match in area density the number of "haloed particles" observed with the electron microscope (Linscott and Hemenway, 1964).

The one exception that we have experienced to the above model occurred in 1973 when we again sampled a NLC in northern Sweden. This time, the flux of residual particles collected appeared to be less than 1/100 the numbers of particles needed to explain the brightness of the cloud we simultaneously sampled. The hydroxyl-ion cluster hypothesis of Witt (1968) may have been at work in this instance. An alternative possibility is that the individual particles we attempted to collect had grown too large for our thin nitrocellulose collections films, and the "iced" particles broke through the collection films. Figure 28 depicts an area of the 1973 collection films showing possible breakthrough holes that appear to be sufficiently numerous to provide the right order of magnitude of the total optical scattering cross section of the NLC. Thus far, in three of the four successful NLC samplings, we have found large enhancements in particle influxes at the time of the NLC activity. Preliminary flux data from one of our 1973 NLC collections are also shown in Figure 27.

Let us next see if the rocket collections are consistent with the zodiacal-light data. Probably the most accurate size distribution of submicron particles, and the one most free from possible collector-efficiency problems, was the nose-cone collection in 1962 (Hemenway, Soberman, and Witt, 1964). Figure 29 shows the integral size distribution per unit area of the particles collected in 1962. The data can be well approximated by a straight line, and a sharp cutoff occurs at about 0.06-μ diameter.

If we consider that the submicron particles have a solar origin, then it would be necessary for them to be of such a nature that solar-radiation pressure and gravity forces would approximately balance. To a first approximation then,

$$r \times \rho = 0.58 \quad , \tag{1}$$

where r is the radius of the particle in microns and ρ the particle's density in g cm^{-3}. Examination of the size distribution cutoff at 0.03-μ radius suggests that the cutoff occurs because nature does not have any materials available with densities greater than about 19 g cm^{-3}. The cutoff at 0.03-μ radius in the NLC sample thus appears to be consistent with a solar origin. The largest particles shown on Figure 29 have a radius of about 0.3 μ and would be expected to have a mass density of about 2 g cm^{-3}. We suspect that the submicron particle population in the solar system has the relationship specified by equation (1). The absence of particles less than 0.03-μ radius is being thoroughly explored with the aid of sample surfaces exposed in the S-149 experiment. We note that, in the solar-facing exposure of S-149, the number density of submicron hole penetrations in thin nitrocellulose film for particles greater than 0.03-μ radius is in fair agreement with the lowest rocket fluxes and the Gemini 9 S-12 data shown in Figure 27.

In order to make a comparison with the zodiacal-light data, we next take the data of Figure 29 and perform the following operations:

1) Convert the integral area distribution of Figure 29 to a differential distribution, keeping track of the differential intervals.

2) Divide the differential distribution by the effective collection time (Soberman and Hemenway, 1965) to obtain the differential flux in the atmosphere. This can be integrated to obtain the integral flux shown for the 1962 NLC collection in Figure 27.

3) Divide by a factor of about 3×10^3 to convert the noctillucent cloud collection of 1962 to the differential flux normally entering the atmosphere, such as that shown in Figure 27 for GZ, October 11, 1972. This is the enhancement factor for the 1962 NLC collection.

4) Divide by another factor of about 10 to convert the differential upper atmospheric fluxes to deep-space differential fluxes to allow for the earth's capture cross section, polar concentration, etc.

5) Divide by an assumed <u>constant</u> velocity of flow outward from the sun, which we take to be 44 km $\sec^{-1}$. Assuming a constant velocity outward is equivalent to saying that the particle motions are such that solar-gravity forces and light pressure balance, and the particles contributing to the zodiacal light obey equation (1).

6) Correct for the differential size interval, and plot the resulting number-density distributions on Figure 30.

Figure 30 is a reproduction of the work of Powell <u>et al.</u> (1967), who attempted to infer from the classic zodiacal-light observations the resulting number-density size distributions at 1 a.u. We have plotted the data from the NLC collections of 1962 for two cases, one assuming a constant mass density (3 g cm^{-3}) for all submicron particles, and the other assuming that the particle density increases as the particle size decreases in accord with equation (1). The fluxes from these two cases agree at about 0.2-μ radius. Note that the variable particle mass-density model agrees somewhat better than the constant mass-density model. The agreement between the zodiacal-light predictions of Powell and our collection data was achieved by assuming that <u>all</u> of the submicron particles are flowing outward from the sun at constant velocity. This suggests that the zodiacal light is primarily a result of submicron particles flowing outward from the sun, rather than of larger cometary debris spiraling inward. The larger particles spiraling inward are the particles needed for the gegenschein. The change in the slope of the meteoroid mass-flux curve at about 10^{-7} g occurs because at masses smaller than this, the particles begin spiraling past the earth in 1 year, and the earth's collection efficiency is thereby reduced.

It is of interest that, if one takes a constant density of 3 g cm^{-3} for the submicron particles and divides the differential flux of the 1962 NLC by about 3×10^5, the outward-flowing mass flux is about 1.2×10^{-14} g m^{-2} sec^{-1}, which represents a total outward mass flow of about 100 tons sec^{-1} through the cross section of the zodiacal-light complex at 1 a.u. ($\sim 10^{22}$ m^2).

If the submicron particles are coming to us from the sun, they would have been formed in a reducing environment and might react chemically in the oxydizing environment of the earth's atmosphere. Evidence of particles reacting in the atmosphere has been seen, and all rocket samples as well as the S-149 samples are kept in an argon environment.

We next explore the possibility that solar particles are able to leave the surface of the sun without evaporating. Sunspots seem to be the only region of the sun's atmosphere that can provide sufficiently low temperatures for particle nucleation and growth. Simple estimates suggest that materials such as HfC (melting point: 4163 K), TaC (melting point: 4153 K), and W (melting point: 3583 K) can form particles within the cooler parts of sunspots (T < 3000 K). It is suspected that the solar abundance of molecular compounds of the refractory elements have been seriously underestimated by solar−spectral studies.

The mass dm of a spherical shell evaporated in a time dt is

$$dm = 4\pi r^2 \rho \, dr = \omega_p \, 4\pi r^2 \, dt \quad , \tag{2}$$

where ρ, m, and r are the particle density, mass, and radius, ω_p is the evaporation rate in cm^{-2} sec^{-1} of the particle, and t is the time in seconds. Integrating (2),

$$t = \frac{\rho r}{\omega_p} \quad . \tag{3}$$

Unfortunately, we do not know the evaporation rates of the solar particles. However, Waldmeier has observed (Bray and Loughhead, 1965, p. 65) faint blue rings surrounding sunspots and extending outside the penumbra of sunspots, about 0.72 times their radius. We suggest that Waldmeier's blue rings represent excess photon scattering in the direction toward the earth by submicron particles carried outward from the umbra of the sunspots by the Evershed effect. The distance along the solar surface that the particles can be carried before they disappear provides a measure of the evaporation rates of the material of the solar particle.

We assume the evaporation rate vs. temperature curve for tungsten and that for the particle to have the same shape but different magnitudes. We rewrite equation (3),

$$t = k_w \frac{\rho r}{\omega_w} \quad , \qquad\qquad (4)$$

where k_w is a constant we must use in order to use also the evaporation rate of tungsten ω_w in place of the evaporation rate of the particle ω_p. We take the sunspot radius to be 15 arcsec, estimate the average tangential velocity of the particle from the Evershed effect to be 0.75 km sec^{-1}, assume the density of the particle to be 20 g cm^{-3}, and the particle radius to be 0.1 μ. We estimate the temperature of a 0.1-μ blackbody to be about 4860 K above the photosphere and the evaporation rate of tungsten at that temperature to be about 7.0×10^{-2} cm^{-2} sec^{-1}. Using these data, we find

$$\frac{15 \text{ arcsec} \times 725 \text{ km arcsec}^{-1}}{0.75 \text{ km sec}^{-1}} = 14,500 \text{ sec} = t = k_w \frac{\rho r}{\omega_w} = k_w \frac{20 \times 10^{-5}}{7.0 \times 10^{-2}} \quad ,$$

$$k_w = 5 \times 10^6 \quad .$$

Note that the value of k_w is very large. This implies a very low evaporation rate for the solar particle at the equilibrium temperature of a 0.1-μ blackbody above the solar photosphere. Since the particles within the solar plasma are probably charged negatively by electron accretion, we suspect that the large value of the k_w means that most of the atoms or molecules of the solar particle become ionized after evaporation and are drawn back into the solar particle. The small fraction of the molecules that are able to escape determines the low evaporation rate of the material of the solar particles.

Table 2 shows the results of integrating outward from the sun to estimate fractions of the mass lost by solar particles leaving the region above a sunspot and the equivalent calculation from a region above the photosphere. The particle survives when leaving the region above a sunspot, but does not after leaving the region over the photosphere. Similar calculations are shown for molybdenum and technetium, with much larger values of k_{Mo} and k_{Tc}, and give about the same answers as did tungsten for the case including coronal heating. This suggests that the fractional mass loss results are independent of the reference material.

The velocities shown in Table 2 are the range of velocities to be expected if solar surges and eruptive phenomena in the vicinity of sunspots provide the initial momentum for the particle ejection, after which only those particles with the velocities specified by equation (1) (or its equivalent including coulomb drag and other outward forces) are able to drift outward through the gravitational trap of the sun and contribute to zodiacal light. Although these preliminary calculations are encouraging, much more remains to be done concerning particle formation and escape from the sun. Also, we must seek additional evidence of the presence of high-temperature refractory materials in the submicron component of the cosmic dust entering the earth's atmosphere.

Collections of cosmic dust and their study are likely to have increased importance in the future for both scientific and practical reasons.

It is noted that Greenberg (1968) has suggested that interstellar particles may have a core-mantle structure. We have made some preliminary estimates that suggest that the nuclei that form interstellar grains may be ejected from sunspots in G-type and cooler stars. Since the more numerous cooler stars would be expected to eject silicon and more abundant elements in their "stardust," it is not surprising that the probable cometary materials shown in Figures 12 and 13 appear to be lower density materials than the "stardust" from our sun. The formation of interstellar-grain nuclei within stellar sun spots may provide a solution to the nucleation problem for interstellar grains. Studies of cosmic dust are likely to become increasingly important to astrophysics and astronomy.

For many years, correlations have been indicated between sunspot cycles and meteorological phenomena. A recent conference at the Goddard Space Flight Center (1973) appears to have concluded that, although there were significant correlations between solar phenomena and weather phenomena, no physical basis for the correlation was found. Our noctilucent-cloud observations suggest that the incoming submicron dust, which we believe to be of solar origin, does nucleate moisture at the 84-km level to form noctilucent clouds. These submicron particles may also nucleate in the stratosphere to form nacreous clouds and may be providing a variable nucleating flux at lower altitudes. It is noted that the fall time through the atmosphere, since the particles are small compared to the mean free path, is independent of size so long as equation (1) is satisfied. The flux of variable-density, submicron cosmic dust might be the missing parameter that makes it difficult for weather forecasts to be made more than several days in the future. It will be important to look for submicron cosmic dust in rainfall.

ACKNOWLEDGMENTS

I am indebted to the National Aeronautics and Space Administration for NASA Grant NGL-33-011-001 and NASA Contract NAS9-10380, which have made this work possible and, particularly, to M. Dubin, whose encouragement and advice over many years has been most helpful.

REFERENCES

Bandeen, W. R., and Maran, S. P., eds., 1973. Symposium on Possible Relationships Between Solar Activity and Meteorological Phenomena, Goddard Space Flight Center, Nov. 7-8.

Berg, O. E., and Gerloff, U., 1971. In Space Research XI, ed. by K. Ya. Kondratyev, M. J. Rycroft, and C. Sagan (Akademie-Verlag, Berlin), p. 225.

Bray, R. J., and Loughhead, R. E., 1965. Sunspots (John Wiley & Sons, New York), 303 pp.

Dubin, M., 1973. COSPAR (in press).

Greenberg, M. J., 1968. In Nebulae and Interstellar Matter VII, (Univ. of Chicago Press, Chicago), p. 221.

Hallgren, D. S., Hemenway, C. L., Mohnen, V. A., and Tackett, C. D., 1973. In Space Research XIII, ed. by M. J. Rycroft and S. K. Runcorn (Akademie-Verlag, Berlin), p. 1099.

Hemenway, C. L., and Hallgren, D. S., 1969. In Space Research IX, ed. by K. S. W. Champion, P. A. Smith, and R. L. Smith-Rose (North-Holland Publ. Co., Amsterdam), p. 140.

Hemenway, C. L., and Hallgren, D. S., 1970. In Space Research X, ed. by T. M. Donahue, P. A. Smith, and L. Thomas (North-Holland Publ. Co., Amsterdam), p. 272.

Hemenway, C. L., Hallgren, D. S., and Coon, R. E., 1967. In Space Research VII, ed. by R. L. Smith-Rose, S. A. Bowhill, and J. W. King (North-Holland Publ. Co., Amsterdam), p. 1423.

Hemenway, C. L., Hallgren, D. S., and Kerridge, J. F., 1968. In Space Research VIII, ed. by A. P. Mitra, L. G. Jacchia, and W. S. Newman (North-Holland Publ. Co., Amsterdam), p. 521.

Hemenway, C. L., Hallgren, D. S., and Schmalberger, D. C., 1972. Nature $\underline{238}$, 256.

Hemenway, C. L., and Soberman, R. K., 1962. Astron. Journ. $\underline{67}$, 256.

Hemenway, C. L., Soberman, R. K., and Witt, G., 1964. Tellus XVI, 84.

Hemenway, C. L., Hallgren, D. S., Laudate, A. T., Patashnick, H., Renzema, T. S., and Griffith, O. K., 1971. In Space Research XI, ed. by K. Ya. Kondratyev, M. J. Rycroft, and C. Sagan (Akademie-Verlag, Berlin), p. 393.

Lindblad, B. A., Arinder, G., and Wiesel, T., 1973. In Space Research XIII, ed. by M. J. Rycroft and S. K. Runcorn (Akademie-Verlag, Berlin), p. 1113.

Link, F., 1973. In Space Research XIII, ed. by M. J. Rycroft and S. K. Runcorn (Akademie-Verlag, Berlin), 1135.

Linscott, J., and Hemenway, C. L., 1964. Tellus XVI, 110.

Patashnick, H., and Hemenway, C. L., 1969. Rev. Sci. Instr. $\underline{40}$, 1008.

Powell, R. S., Woodsen, P. E. III, Alexander, M. A., Circle, R. R., Konheim, A. G., Vogel, D. C., and McElfresh, T. W., 1967. ZLIM NASA SP-150, 225.

Rauser, P., and Fechtig, H., 1972. In Space Research XII, ed. by S. A. Bowhill, L. D. Jaffe, and M. J. Rycroft (Akademie-Verlag, Berlin), p. 391.

Rauser, P., and Fechtig, H., 1973. In Space Research XIII, ed. by M. J. Rycroft and S. K. Runcorn (Akademie-Verlag, Berlin), p. 1127.

Roach, F. E., Pettit, H. B., Tandberg-Hanssen, E., and Davis, D. N., 1954. Astrophys. Journ. $\underline{119}$, 253.

Rosen, J. M., 1969. Space Sci. Rev. $\underline{9}$, 58.

Soberman, R. K., and Hemenway, C. L., 1965. Journ. Geophys. Res. $\underline{70}$, 4943.

Whipple, F. L., 1950. Proc. Nat. Acad. Sci. $\underline{36}$, 687.

Witt, G., 1968. In Space Research XIV, ed. by A. P. Mitra, L. G. Jacchia, and W. S. Newman (North-Holland Publ. Co., Amsterdam), p. 157.

TABLE 1.

Flux-altitude data (noctilucent cloud display, July 31, 1971).

Altitude interval (km)	Average particle diameter interval (μ)	Differential flux (no. m^{-2} sec^{-1})	Integrated flux (no. m^{-2} sec^{-1})
79 - 90.4	0.1 - 0.2	1.1×10^3	2.2×10^3
	0.2 - 0.3	1.9×10^2	1.1×10^3
90.9 - 101.3	0.1 - 0.2	2.9×10^2	1.1×10^3
	0.2 - 0.3	3.6×10^2	7.7×10^2
101.7 - 113.5	0.1 - 0.2	5.3×10^3	1.0×10^4
	0.2 - 0.3	3.4×10^3	4.7×10^3

TABLE 2.

Estimated fractional mass losses of 0.1-μ radius blackbody
particles leaving the sun.

Velocity (km sec^{-1})	Sunspot (no coronal heating)	Photosphere	Sunspot	Sunspot (with coronal heating)	Sunspot
	(w)	(w)	(w)	(Mo)	(Tc)
20	0.0099	1.87	0.057	0.014	0.043
30	0.0066	1.25	0.038	0.0091	0.028
50	0.0040	0.75	0.023	0.0054	0.017
100	0.0020	0.375	0.011	0.0030	0.0095
200	0.00099	0.187	0.006	0.0014	0.0043

$$k_W = 5.08 \times 10^6 \qquad k_{Mo} = 1.28 \times 10^8 \qquad k_{Tc} = 9.9 \times 10^{12}$$

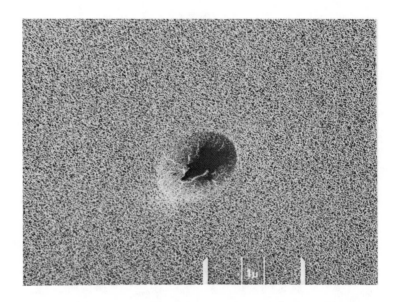

Figure 1.   Electron micrograph of double-component particle.

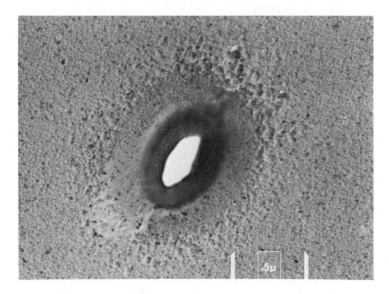

Figure 2.   "Evil eye" from Gemini 9 S-12 experiment.

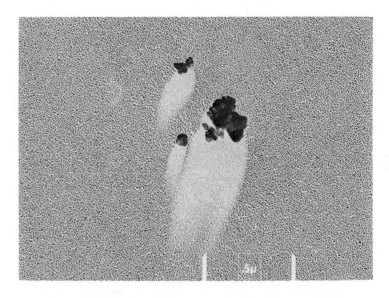

Figure 3.   Electron micrograph of cluster of particles.

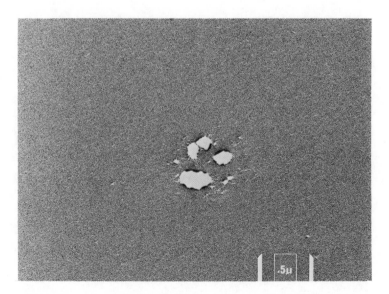

Figure 4.   Cluster of holes from Gemini 9 S-12 experiment.

Figure 5. Electron micrograph of a single large particle plus smaller particles.

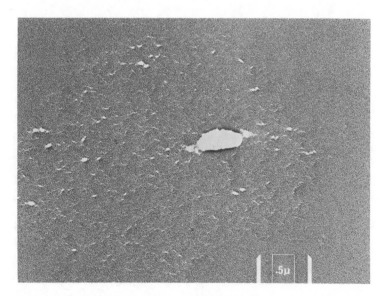

Figure 6. Cluster of holes from Gemini 9 S-12 experiment.

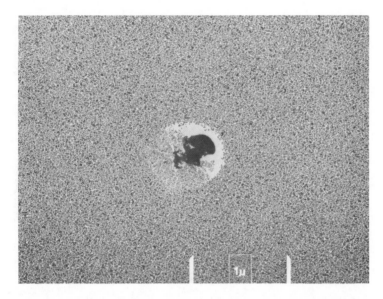

Figure 7.   Electron micrograph of double-component particle.

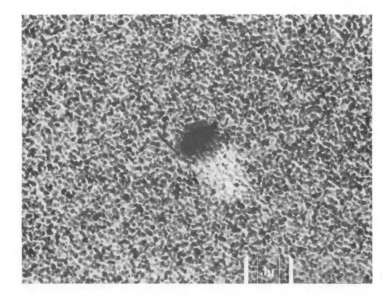

Figure 8.   Electron micrograph of rocket collection particle.

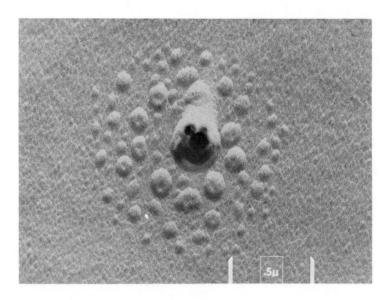

Figure 9. Electron micrograph of double-component particle from 1962
           noctilucent cloud experiment.

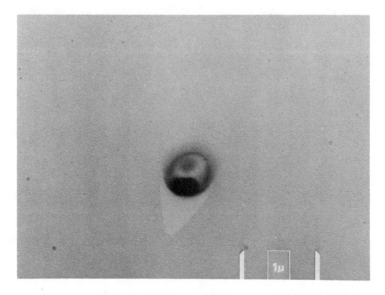

Figure 10. Electron micrograph of rocket collection particle.

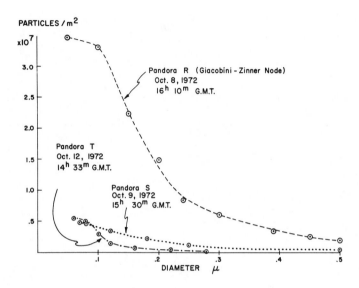

Figure 11.   Cumulative size distributions from Giacobini Zinner, 1972 and after.

Figure 12.   Electron micrograph of conglomerate particle.

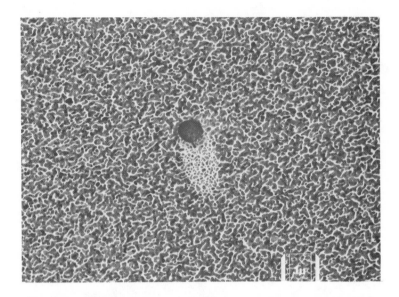

Figure 13. Electron micrograph of rocket collection particle.

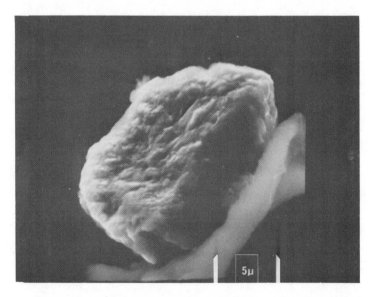

Figure 14. Scanning electron micrograph of particle from balloon collection.

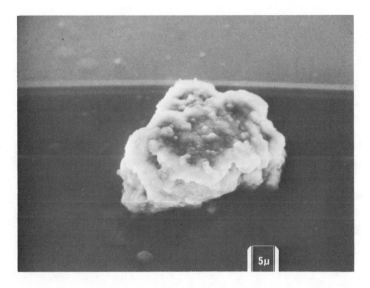

Figure 15.   Scanning electron micrograph of particle from balloon collection.

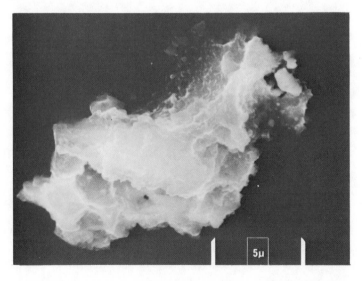

Figure 16.   Scanning electron micrograph of particle from balloon collection.

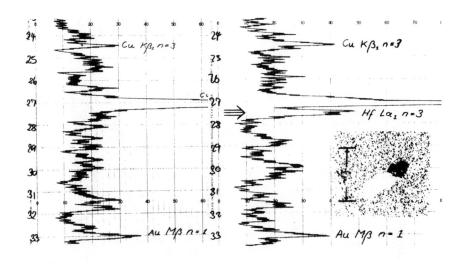

Figure 17. X-ray spectra of submicron cosmic dust particles and control area.

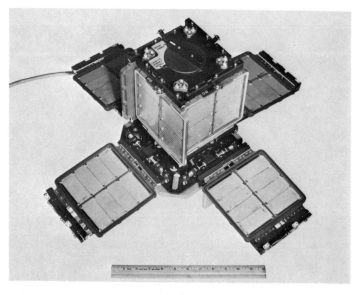

Figure 18. Skylab S-149 micrometeorite detection experiment.

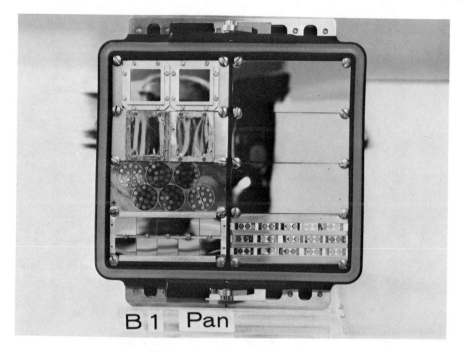

Figure 19.   S-149 samples.

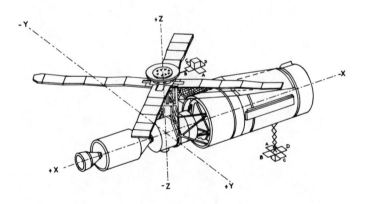

Figure 20.   S-149 Skylab exposure locations (+Z toward sun).

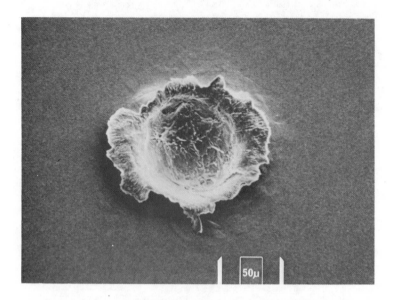

Figure 21. Impact crater from S-149 experiment.

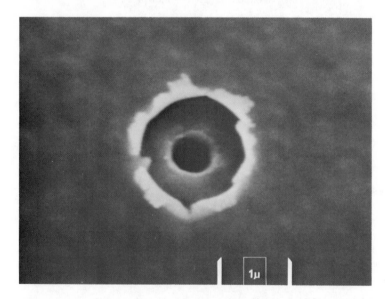

Figure 22. Impact crater on copper covered with gold foil.

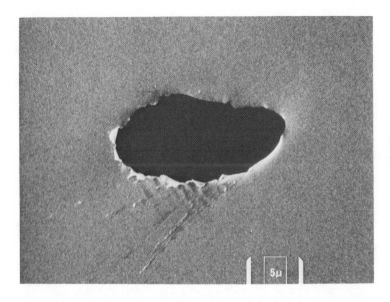

Figure 23.   Penetration hole in gold film from S-149 experiment.

Figure 24.   Cluster of holes in gold film located directly beneath hole shown in
             Figure 23.

Figure 25. Solar-facing silver and stainless steel S-149 slides.

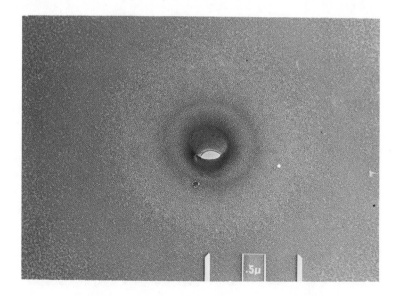

Figure 26. "Evil eye" from S-149 experiment.

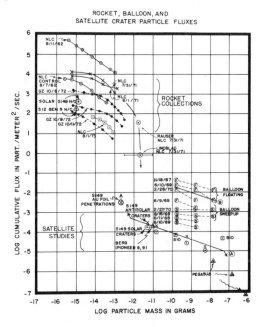

Figure 27.   Rocket, balloon, and satellite crater particle fluxes.

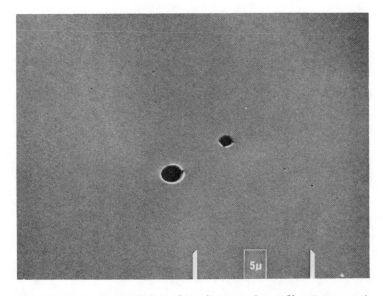

Figure 28.   Holes in nitrocellulose film from rocket collection experiment.

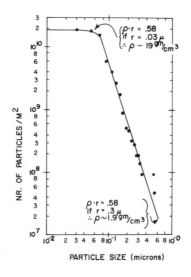

Figure 29.   Cumulative area size distribution, 1962 noctilucent cloud experiment.

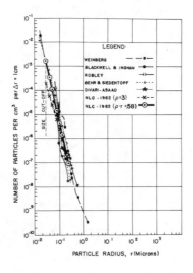

Figure 30.   Comparison of cosmic-dust size distribution predictions from
zodiacal light data with collection size distribution.

THE FINE-GRAINED STRUCTURE OF CHONDRITIC METEORITES

John A. Wood

Center for Astrophysics
Harvard College Observatory and Smithsonian Astrophysical Observatory
Cambridge, Massachusetts

ABSTRACT

Type I and II carbonaceous chondrites consist largely of fine-grained mixtures of types of minerals that would be thermodynamically stable in the solar nebula only at relatively low temperatures (< 500°K). These include hydrated Mg silicates, Fe_3O_4, and carbonate and sulfate compounds. Their aggregate composition is similar to that of solar matter, so far as proportions of metallic elements is concerned. This low-temperature material may have condensed originally as high-temperature minerals dispersed in the nebula, after which reaction with the cooling nebular gases converted the mineral grains to phases stable at lower temperatures; but considerations of crystal morphology and mineral chemistry make it seem likely that the currently observable low-temperature minerals condensed directly from the vapor phase.

1. INTRODUCTION

It is commonly held that the planetary system formed from a gaseous nebula that surrounded the infant sun as the latter evolved toward the main sequence. Condensation within the nebula as it cooled produced grains of minerals, metals, and ices that then accreted, by a poorly understood process, into planets and their satellites. In those bodies we have been able to study (earth and moon), geologic evolution has totally destroyed the original assemblages of nebular condensate grains, reconstituting their substance into new and different types of rock. However, the original agglomeration of condensate grains is believed to be preserved in certain classes of meteorites. The most persuasive evidence for the primitive character of these meteorites is their unfractionated pattern of elemental abundances, which closely parallels that of the solar photosphere (MacDonald, 1959; Ringwood, 1961). Further, the fine-grained and chaotic nature of their mineralogy (discussed below) is consistent with their formation by accumulation from a system of dispersed dust particles. Presumably, each of the tiny mineral grains or entities of other sorts was once an isolated particle moving freely in the primordial solar nebula. Presumably also, these masses of agglomerated material never joined planets or satellites of any substantial size, and for this reason were not subsequently reconstituted by the processes of melting and chemical differentiation that have affected terrestrial and lunar rocks. (The total amount of geological activity experienced by a planet appears to be a function of its size.) These meteorites may have existed since the origin of the solar system in small asteroids, or, possibly, they are representative of the nonvolatile component of comets. By studying them, we are able to learn something about the character of the dust particles generated as a by-product of star formation.

The type of primitive unevolved meteorites I refer to are called carbonaceous chondrites. They comprise about 4% — by number — of the meteorites observed to fall to earth. (This is by no means an accurate measure of their abundance in the solar system, however. Measurements of the spectral reflectivities of asteroid surfaces and their comparison with spectral reflectivities of various meteorite types measured in the laboratory indicate that carbonaceous chondrite material is far more abundant in the asteroid belt than it is in our meteorite collections [McCord and Gaffey, 1974]. One factor that biases the statistics of our meteorite collections against carbonaceous chondrites is the relative friability or mechanical weakness of the latter; carbonaceous chondrites are much less likely to survive deceleration and flight through the terrestrial atmosphere than are the other known meteorite classes.)

There are three principal subclasses of carbonaceous chondrites, termed Type I, Type II, and Type III, or C1, C2, and C3. Chemical compositions of representatives of each subclass appear in Table 1. The most conspicuous chemical differences between the subclasses appear in their content of the elements incorporated in the relatively volatile compounds: H (in H_2O), C (usually as complex organic compounds), and S (sulfate and sulfide minerals). The content of volatile materials is taken to be an index of the "primitiveness" of a chondrite; abundant volatiles in a chondrite subclass appear to imply that the nebular gases had fallen to quite low temperatures before the grains they had condensed were accreted into this particular mass (thereby sealing off the system from the addition of material that might condense at even lower temperatures). Some chondrite types — the so-called ordinary chondrites — appear to have accreted and become closed systems at relatively high temperatures; abundances of a large number of relatively volatile elements are systematically depleted in these meteorites. For this reason, C1 chondrites are generally held to be the most primitive form of planetary material available to us and the most likely to represent a straightforward accumulation of dust grains that condensed in the primordial solar nebula. C3 chondrites are least interesting from this point of view: They contain abundant chondrules, solidified droplets that appear to represent the product of some process of remelting that affected the original solar system condensates; in addition, some C3 chondrites show mineralogical evidence of having been heated and, to some extent, metamorphosed after they had accreted into their present configuration.

Specimens of all three carbonaceous chondrite subclasses are shown in Figure 1. My discussion will focus on the principal fine-grained minerals in C1 and C2 chondrites; these mineral grains presumably comprised the bulk of the dust particles that condensed in the vicinity of the asteroid belt as the solar nebula cooled. I will not attempt to enumerate and discuss all minor minerals in these meteorites. The fine-grained mineralogy of the Allende C3 chondrite will also be noted, and some of the problems that prevent a rationalization of the mineral constituents of carbonaceous chondrites in terms of a straightforward model involving condensation from a gas of solar composition will be discussed.

2. LAYER–LATTICE SILICATES

The great majority of fine mineral grains in Cl and C2 chondrites can be termed layer–lattice silicates. The name is descriptive of their crystallographic structure, which consists of parallel sheets or layers of joined SiO_4 tetrahedra. The individual sheets are loosely bonded together by hydroxyl ions and metallic cations (Figure 2). Members of this mineralogic family, which includes the mica and clay minerals, tend to split easily along their layer planes and thus are flaky in character. The appearance of layer–lattice silicate minerals in carbonaceous chondrites is shown in Figures 3 and 4. X-ray studies of these minerals have indicated variously that they belong to the subfamily of chlorites, having the general formula $(Mg, Al, Fe)_{12} ((Si, Al)_8 O_{20}) (OH)_{16}$ (Nagy, Meinschein, and Hennessy, 1963; Boström and Fredriksson, 1966), or to the subfamilies serpentine, $Mg_3 (Si_2 O_5) (OH)_4$, and montmorillonite, $(\frac{1}{2} Ca, Na)_{0.7} (Al, Mg, Fe)_4$ $(Si, Al)_8 O_{20} (OH)_4 \cdot nH_2 O$ (Mason, 1963; Bass, 1971). Electron microprobe analyses of the layer–lattice silicates in a Cl and a C2 chondrite are shown in Table 2. Extensive elemental substitutions are permitted in the layer–lattice silicate minerals, as indicated by the formulas just given. They can be said to be "forgiving" minerals which, during condensation of a gas of complex composition, might be expected to incorporate nonpreferentially practically all the metallic atoms present in the gas. It is interesting to compare the composition of these chondritic layer–lattice silicates with the cosmic abundance of the most abundant metallic elements (Table 2): It can be seen that with the exception of S (which is not a metallic element) and Na (which was not analyzed in either of the samples reported), the abundances of elements incorporated in these meteoritic minerals are remarkably similar to the cosmic abundance pattern.

3. MAGNETITE

Although Table 2 shows that layer–lattice silicates, which comprise the bulk of Cl and C2 carbonaceous chondrites, are capable of accommodating Fe in cosmic proportions, Fe also appears in these meteorites in the form of magnetite ($Fe_3 O_4$). The abundance of the magnetite is only a few percent, however – an order of magnitude less than the abundances of layer–lattice minerals. Although some magnetite in some C2 chondrites occurs as a layer coating grains of metallic nickel-iron and was clearly produced as a "corrosion" product by reaction of the metallic Fe with its environment at some stage, most of the magnetite in C1 and C2 chondrites is not associated with metal. It often occurs in curious and very regular morphologies (Figure 4), which would be difficult to rationalize by any mechanism of formation other than condensation of the magnetite directly from the vapor state. The Ni content of these magnetite plates and polyhedra is very low, $\sim 0.1\%$. This further militates against the possibility that the particles were formed by oxidation of earlier formed metal in a cooling nebular environment: Primary chondritic metal invariably contains 5% or more Ni, and there is no way in which this could be physically removed from a metal particle during its oxidation. The Ni in oxidation-formed magnetite would still be present, either in solution in the magnetite crystal or possibly as finely disseminated unreacted metal.

It is very probable that the magnetite plates and polyhedra formed by direct condensation from the solar nebula. The difference in their morphologies probably reflects a difference in the physical circumstances of condensation. All other things being equal, crystals that precipitate from supercooled systems tend to have less equidimensional morphologies (i.e., are more platey or rod-like in form) than do crystals of the same compound that form just beneath their equilibrium crystallization temperature.

4. SULFUR-BEARING MINERALS, CARBON, AND CALCIUM

Sulfur, C, and possibly Ca are the only elements present in major amounts in carbonaceous chondrites that cannot be accommodated in the layer-lattice silicate minerals or magnetite. In C1 chondrites, S is present primarily in the form of sulfate minerals ($CaSO_4 \cdot 2H_2O$ (gypsum), $MgSO_4 \cdot nH_2O$), secondarily as troilite (FeS). Carbon is present partly in the form of carbonate minerals and graphite, but chiefly in the form of an extremely complex array of organic compounds, which appear to have formed by abiotic reactions that occurred in the cooling solar nebula before accumulation (Anders, Ryoichi, and Studier, 1973). The carbonaceous compounds occur dispersed throughout the matrix of C1 and C2 chondrites, presumably coating the individual mineral grains, and are not concentrated in discrete masses. Most of the Ca in C1 chondrites appears in the form of gypsum. The sulfate and carbonate minerals in C1 chondrites are water soluble; in part, these minerals appear in the form of veins filling cracks in C1 chondrites (Figure 1a), indicating that they were dissolved and reprecipitated by liquid water that flowed through the parent mass of the C1 chondrites at some time after its accretion.

Sulfur occurs in the matrices of C2 chondrites as troilite and pentlandite ((Fe, Ni)S), but also in the form of an enigmatic mineral (Figure 5a) having the color of tarnished brass, which has not yet been sufficiently well characterized to be recognized as a discrete mineral species and given a name. An analysis of this complex S-bearing mineral appears in Table 3. Much of the C in the matrices of C2 chondrites is present as complex organic compounds. Additional C and much of the Ca in these chondrites occurs in a form of calcite ($CaCO_3$).

5. MATRICES OF C3 CHONDRITES

Except for the S compounds troilite and pentlandite, the minerals described above, which comprise the bulk of C1 chondrites and the matrices of C2 chondrites, are scarce or totally lacking in C3 chondrites. The matrix of the Allende C3 chondrite consists largely of olivine ((Fe, $Mg)_2SiO_4$), in which $Fe/Mg \simeq 1$. The olivine occurs in well-formed crystals of small (~ 1 μ) and remarkably uniform dimensions (Figure 5b).

Some C3 chondrites contain unequivocal evidence, in the form of element distributions and textures in metallic minerals (Wood, 1967), of having been processed or metamorphosed at high temperatures (~ 800 K) subsequent to the time of their accretion. Temperatures this high would have the effect of dehydrating layer-lattice silicate minerals and transforming them to anhydrous minerals, chiefly olivine. However, it is unlikely that the olivine crystals in the matrix of Allende were formed in this way. Their crystal morphologies are characteristic of independently precipitated crystals, and dissimilar to an assemblage of grains that crystallized while pressing in on one another. Further, radiation damage preserved in the chondrules of Allende indicate that this particular chondrite has not been appreciably heated since the time of its accretion (Green, 1971). It appears that the principal matrix mineral of Allende condensed at a relatively high temperature (>500 K) and was removed from contact with the solar nebula by accretion into a chondritic mass before the temperature of the system had fallen to the lower values (<400 K) at which reaction with the nebular gases would transform olivine into layer-lattice silicates.

6. FORMATION OF THE FINE-GRAINED MINERALS IN CARBONACEOUS CHONDRITES BY CONDENSATION

There is no pressure and temperature regime in which layer-lattice silicate minerals and magnetite, the chief minerals of C1 and C2 chondrites, could form stably by condensation directly from a gas containing the cosmic proportions of elements (Figure 6). In all cases, olivine and metallic iron should have condensed at higher temperatures, and layer-lattice silicates and magnetite should form subsequently by reaction of the residual gas with olivine and metal particles at lower temperatures. However, most of the magnetite in C1 and C2 chondrites does not seem to have formed in this way. The morphologies of magnetite crystals shown in Figures 4a, b, c, and their low content of Ni, strongly indicate that they precipitated directly from the gas phase. It is difficult to say whether the layer-lattice silicate minerals formed by direct condensation or by the transformation of preexisting olivine grains; the complex chemical compositions of the layer-lattice silicates and their similarity to the cosmic pattern of elemental abundances (Table 2) suggest that this mineral, too, condensed directly from the gaseous nebula, since olivine has a rigorously defined chemical composition and cannot have accommodated itself to the proportions of elements shown in Table 2. (The strength of this argument is somewhat weakened, however, by the fact that one cannot be sure that analyses of pure, uncontaminated layer-lattice silicates are reported in Table 2. The layer-lattice silicates in carbonaceous chondrites are extremely fine grained, much smaller in grain size than the diameter of an electron microprobe beam; the possibility cannot be ruled out that submicroscopic grains of other minerals, having substantially different chemical compositions, are interleaved between the layer-lattice silicate flakes and were included in the analyses.)

The condensation sequences for C1, C2, and C3 chondrites are briefly summarized in Table 4. It appears that C1 chondrites are the product of low-temperature condensation only; C3 chondrites were produced by high-temperature condensation; and C2 chondrites contain condensation products from the entire range of possible temperatures.

7. CHONDRULES

The fine-grained mineral constituents described above are those that appear to have condensed directly from the vapor to the solid state. In addition to these, C2 and C3 chondrites contain rounded entities (chondrules) that clearly were at least partly molten at some time in the past (Figure 7a). In many cases, part or all of the formerly melted silicate material is preserved in the form of glass. The origin of these erstwhile liquid droplets is puzzling. Earlier I argued that they too represent solar system condensates (Wood, 1963), formed under conditions where condensation directly from the gas to the liquid state was possible. However, extraordinary physical conditions would be required to accomplish this. In order to condense magnesium silicate liquids from a gas containing the cosmic abundances of the elements, total gas pressures in excess of 100 atm are required. Even when allowance is made for the fact that chondrules contain other metallic elements in addition to Mg and Si, and that many of the chondrules need never have been completely melted, the minimum pressure requirement for condensation of liquids adequate to reproduce the properties of the chondrules appears to be several atmospheres. Maximum pressures attainable in model solar nebulae, by comparison, appear to be only about 10^{-3} atm (Cameron and Pine, 1973).

Alternatively, the chondrules can be understood as having been formed by remelting of agglomerations of silicate grains that originally condensed directly into the solid state. This remelting might have been promoted by unspecified high-energy events in the solar nebula, before the dust aggregations had accreted into planets, or by high-energy events in or on the parent chondrite planets after their accretion. Recently, much interest has been shown in the possibility of understanding the chondrules as melt droplets from impact cratering events on the surfaces of the parent chondrite planets, and attention has been drawn to the "chondrules" found in soil samples returned from the lunar surface, which has been bombarded by meteorites through all of geologic time (King, Butler, and Carman, 1972; Nelen, Noonan, and Fredriksson, 1972). It seems to me that the lunar analogy weakens the case for impact-produced chondrules on asteroidal surfaces, rather than supporting it: "Chondrules" are in fact very rare in the lunar soils (we found less than 1% of spherules among lunar soil particles in the 1- to 2-mm size range examined by our group at the Center for Astrophysics. The soils consists largely of broken rather than melted debris. Where the meteorite bombardment has caused melting on the lunar surface, the product almost invariably takes the form of irregular, untidy masses of mixed glass and unmelted soil debris (Figure 7b). Many of the chondritic meteorites, on the other hand, contain a very high component ($\gtrsim 50\%$) of spheroidal chondrules or fragments of such chondrules.

It is probable that at this point nobody has come very close to understanding what deviations from the pattern of straightforward condensation from a cooling solar nebula gave rise to the chondrules. I have suggested that circumstances in the primordial nebula (such as solid condensation, local concentration of condensates, and revaporization) could give rise to local enhancements of the ratio of abundances of condensable metallic elements and H_2O to hydrogen. The pressure requirement for condensation as a liquid during subsequent cooling in such a region would be reduced. Chemical enhancements of this sort could also make possible the equilibrium condensation of magnetite and layer-lattice silicate minerals, as opposed to their formation from preexisting metallic iron and olivine or their formation by nonequilibrium processes (condensation from a supercooled gas).

REFERENCES

Anders, E., Ryoichi, H., and Studier, M., 1973. Science 182, 781.

Bass, M. N., 1971. Geochim. Cosmochim. Acta 35, 139.

Boström, K., and Fredriksson, K., 1966. Smithsonian Misc. Coll. 151, 53 pp.

Cameron, A. G. W., 1973. Space Sci. Rev. 15, 121.

Cameron, A. G. W., and Pine, M. R., 1973. Icarus 18, 377.

Clarke, R. S., Jr., Jarosewich, E., Mason, B., Nelen, J., Gómez, M., and Hyde, J. R., 1970. Smithsonian Contr. Earth Sci. No. 5, 53 pp.

Fuchs, L., Olsen, E., and Jensen, K. J., 1973. Smithsonian Contr. Earth Sci. No. 10, 39 pp.

Green, H. W., 1971. Science 172, 936.

Grossman, L., 1972. Geochim. Cosmochim. Acta 36, 597.

Jarosewich, E., 1971. Meteoritics 6, 49.

Jedwab, J., 1971. Icarus 15, 319.

King, E. A., Jr., Butler, J. C., and Carman, M. F., 1972. Proc. Third Lunar Sci. Conf., Geochim. Cosmochim. Acta, Suppl. 3, 1, (The M.I.T. Press, Cambridge), p. 673.

Lewis, J. S., 1972. Earth Planet. Sci. Lett. 15, 286.

MacDonald, G. J. F., 1959. Researches in Geochemistry, ed. by P. Abelson, (John Wiley, New York), p. 476.

Mason, B., 1963. Space Sci. Rev. 1, 621.

McCord, T. B., and Gaffey, M. J., 1974. Science 186, 352.

McMurchy, R. C., 1934. Zeits. Krist. 88, 420.

Nagy, B., Meinschein, W. G., and Hennessy, D. J., 1963. Ann. N.Y. Acad. Sci. 108, 534.

Nelen, J., Noonan, A., and Fredriksson, K., 1972. Proc. Third Lunar Sci. Conf., Geochim. Cosmochim. Acta, Suppl. 3, 1 (The M.I.T. Press, Cambridge), p. 723.

Ringwood, A. E., 1961. Geochim. Cosmochim. Acta 24, 159.

Whipple, F. L., 1966. Science 153, 54.

Wiik, H. B., 1956. Geochim. Cosmochim. Acta 9, 279.

Wood, J. A., 1963. Icarus 2, 152.

Wood, J. A., 1967. Icarus 6, 1.

Wood, J. A., Marvin, U. B., Reid, J. B., Jr., Taylor, G. J., Bower, J. F., Powell, B. N., and Dickey, J. S. Jr., 1971. Smithsonian Astrophys. Obs. Spec. Rep. No. 333, 272 pp.

TABLE 1.

Chemical compositions (weight percents) of representative
C1, C2, and C3 chondrites.

	Orgueil (C1)[1]	Murchison (C2)[2]	Allende (C3)[3]
SiO_2	22.59	29.07	34.26
TiO_2	0.07	0.13	0.14
Al_2O_3	1.65	2.15	3.18
Cr_2O_3	0.35	0.48	0.53
FeO	23.72	22.39	27.09
MnO	0.19	0.20	0.18
MgO	15.81	19.94	24.75
CaO	1.22	1.89	2.57
Na_2O	0.74	0.24	0.45
K_2O	0.07	0.04	0.03
P_2O_5	0.27	0.23	0.23
$H_2O(+)$	20.08	8.95	<0.1
$H_2O(-)$		1.14	0.00
SO_3	13.70	0.90	—
C	3.10	1.85	0.29
CO_2	—	1.00	—
Fe	—	0.13	0.19
Ni	—	—	0.40
NiO	1.27	1.75	—
NiS	—	—	1.56
FeS	—	7.24	4.08
Sum	104.83	99.72	99.93

[1] Wiik (1956). Calculated so all H appears as H_2O, all S as SO_3, all C as elemental C. A more realistic association of elements would require less O, resulting in a sum closer to 100.0%.

[2] Jarosewich (1971).

[3] Clarke et al. (1970).

TABLE 2.

Proportions of major condensable elements in layer-lattice silicates
of C1 and C2 chondrites, and in the table of cosmic abundances.

(Atomic percents; normalized to Si = 1.00)

Element	Orgueil (C1)[1]	Murchison (C2)[2]	Cosmic abundance[3]
Mg	0.87	0.90	1.06
(Si	1.00	1.00	1.00)
Fe	1.05	1.21	0.83
S	0.18[4]	~0	0.50
Al	0.07	0.15	0.09
Ca	N.A.[5]	0.04	0.07
Na	N.A.	N.A.	0.06
Ni	0.09	0.05	0.05
Mn	N.A.	0.01	0.01
Cr	N.A.	0.01	0.01

[1] Boström and Fredriksson (1966).

[2] Fuchs, Olsen, and Jensen (1973).

[3] Cameron (1973).

[4] Probably in sulfate contaminant.

[5] N.A. = not analyzed.

TABLE 3.

Composition of poorly characterized S phase in
Murchison C2 chondrite (Fuchs, Olsen, and
Jensen, 1973); weight percents

Element	Percent
Fe	42.0
Ni	6.1
Cr	1.6
S	17.6
O	31.0
C	0.2
P	0.3
	98.8

TABLE 4

Approximate condensation temperatures of minerals in carbonaceous chondrites
(Grossman, 1972; Fuchs, Olsen, and Jensen, 1973).

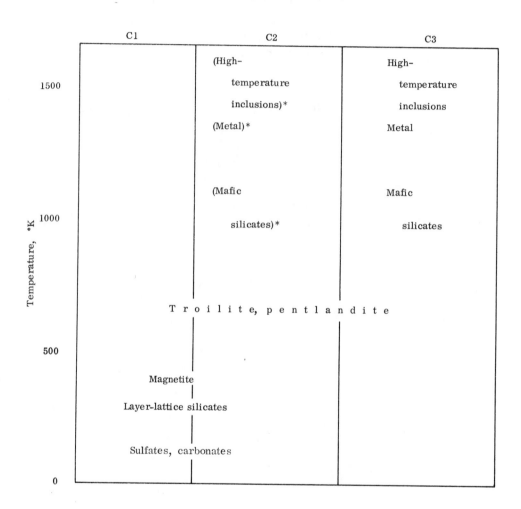

*Brackets denote condensation in relatively small
amounts.

l cm

Figure 1.   Specimens of the three subclasses of carbonaceous chondrites: a)
Orgueil, C1 chondrite; note fusion crust
at top of fragment and light vein of water-
soluble minerals perpendicular to it.
Consists chiefly of layer-lattice silicate
minerals (chlorite or serpentine) and
bituminous organic compounds. Photograph courtesy of U.S. National Museum.
b) Pollen, C2 chondrite, with millimeter
scale. Scattered chondrules and mineral
grains (light color) are visible, but the
chondrite consists largely of matrix
material that is compositionally rather
similar to the substance of C1 chondrites.
c) Allende, C2 chondrite with millimeter
scale. Irregular white inclusions are
aggregations of refractory condensates,
less conspicuous rounded structures are
chondrules; both are embedded in dark
matrix of fine-grained olivine and graphite.

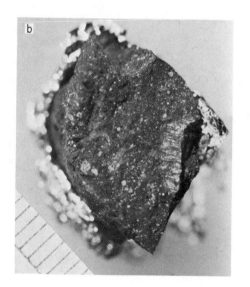

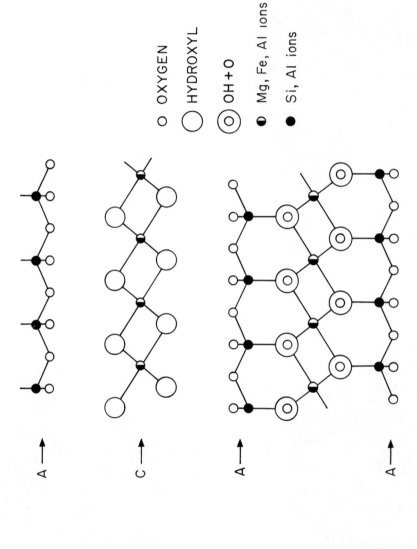

OXYGEN

HYDROXYL

OH + O

Mg, Fe, Al ions

Si, Al ions

Figure 2. Schematic drawing of the structure of chlorite (after McMurchy, 1934), a typical layer-lattice silicate mineral. These are based on sheets (A, viewed edge-on) composed of SiO_4 tetrahedra that share three of their four O atoms with neighboring tetrahedra. The sheets are joined in pairs by Mg^{2+}, Fe^{2+}, Al^{3+}, and OH^- ions (B). The paired sheets are loosely bonded by a structure consisting of these same ions (C). The mineral cleaves easily into flakes along the latter joins.

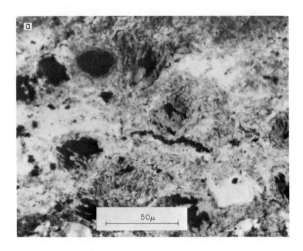

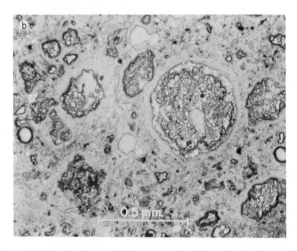

Figure 3.   Textures in the Pollen C2 chondrite.   a) Thin section (~10 μ thick) of the
matrix, by transmitted light.   Dark minerals (opaque) are magnetite,
metal, and sulfides; light minerals are silicates or calcite; the swirly gray
remainder (actually yellowish green in color) consists largely of layer-
lattice silicate minerals, more or less darkened by carbonaceous material.
b) Polished surface, by reflected light, showing rounded chondrules and
irregular inclusions, embedded in fine-grained matrix.   Note tendency of
flaky layer-lattice silicate minerals in latter to pack around and parallel
surfaces of harder silicate chondrules and inclusions.   Largest round
chondrule is rimmed by poorly characterized S phase of Table 3 (lightest
gray.

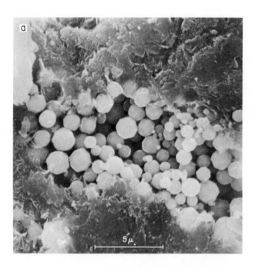

Figure 4.   Morphologies of magnetite and layer-lattice silicate in the Orgueil C1
              chondrite, as viewed by scanning electron microscope (SEM) and trans-
              mission electron microscope (TEM). a) Near-equant crystals of magne-
              tite, nested in a cavity in flaky layer-lattice silicate minerals; SEM.
              Magnetite appears spheroidal by light microscopy; higher resolution SEM
              and TEM images show them to have well-developed cubo-dodecahedral
              crystal faces. b) Aggregation of cubo-dodecahedral magnetite crystals,
              possibly bonded by ferromagnetic attraction; SEM.   This overall form is
              termed a "framboid" by Jedwab (1971).

Figure 4 (cont.)   c) Stacked plates of magnetite; SEM.   Photographs from Jedwab
(1971).   d) Layer-lattice silicate mineral from Orgueil, by TEM.
Photograph from Boström and Fredriksson (1966).

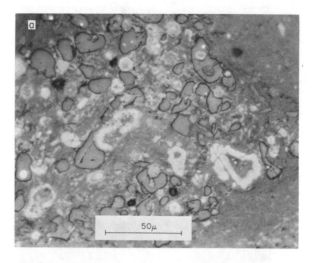

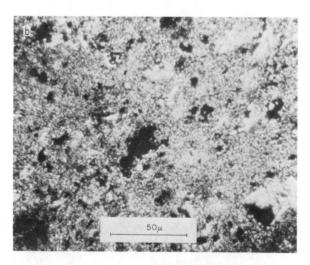

Figure 5.   a) Poorly characterized S phase (white to light gray, forming concentric
             encrustations) in the Pollen C2 chondrite. Image by reflected light; gray
             minerals with well-defined dark outlines are silicates. Gray material
             interstitial to silicates and S phase is matrix, consisting chiefly of layer-
             lattice silicate. b) Matrix of the Allende C3 chondrite, by transmitted
             light. This consists largely of olivine crystals of remarkably uniform
             dimension (~1 μ) and morphology.

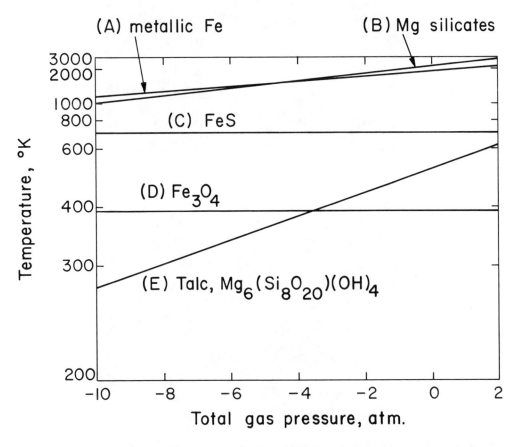

Figure 6. Pressure/temperature plot showing positions of reaction curves along which several important condensations or reactions would occur in a cooling gas system containing the cosmic abundances of elements. Along (A), metallic Fe should condense from the vapor state; along (B), solid Mg_2SiO_4 and $MgSiO_3$ should appear (Grossman, 1972). Metallic Fe should transform to FeS on (C), and to Fe_3O_4 (magnetite) on (D). Magnesium silicates would transform to talc, a layer-lattice silicate mineral, along (E) (Lewis, 1972). Transformation curves for other layer-lattice silicate minerals, such as chlorite or serpentine, should not differ greatly from this in position.

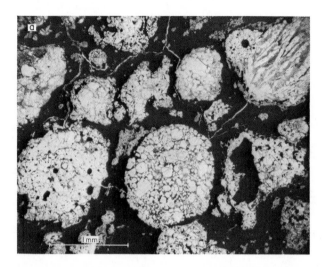

Figure 7.   Melted particles of planetary material. a) Chondrules in the Renazzo C2
             chondrite; thin section, by transmitted light. Chondrules consist largely
             of Mg silicate crystals (white), surrounded by glass (gray). Black matrix
             between chondrules consists of fine-grained layer-lattice silicates and
             magnetite. Figure from Whipple (1966). b) "Glass bonded agglutinates"
             picked from Apollo 12 soil sample 12070; a particularly common type of
             soil particle. Each mass consists of a twisted, bubbly glob of melted lunar
             soil, enclosing an untidy array of unremelted or partly remelted soil grains
             (Wood et al., 1971).

DUST IN THE SOLAR NEBULA

Lawrence Grossman

The University of Chicago, Chicago, Illinois

ABSTRACT

The chemical and mineralogical features of the white, Ca-rich inclusions in Allende and other carbonaceous chondrites are strikingly similar to those predicted from thermodynamic models for the highest temperature condensates from the solar nebula. Many of the physical and chemical properties of the chondritic minerals may thus be quite like those of the dust in interstellar regions. The oxygen isotopic composition of meteoritic condensates suggests that they contain a component of interstellar dust that survived the birth of the solar system.

1. INTRODUCTION

In recent years, equilibrium thermodynamic models have improved our understanding of the sequence of condensation of compounds from a cooling gas of solar composition. This paper will examine the predictions of these models and will compare them with mineralogical features of carbonaceous chondrites. Because interstellar grains probably formed by condensation in cooling, relatively dense stellar nebulae, the study of possible solar nebular condensates in chondrites may have much broader applications than was previously thought. Finally, recent evidence for the direct incorporation of interstellar grains in meteorites will be discussed.

2. THEORY

The basic theory behind the computer calculations on which this discussion is based is explained in detail in Grossman (1972), although some of the results presented here differ from those because newer thermodynamic and abundance data have been used.

If the composition, pressure, and temperature of a system are known, the relative abundances of crystalline phases and concentrations of gaseous species can be calculated from free-energy data. In these calculations, the solar system is assumed to consist of the 15 most abundant elements, excluding the noble gases, in the proportions given by Cameron's (1973) abundance table. The pressure and temperature range investigated is consistent with the Cameron and Pine (1973) hydrodynamic model for the evolution of the solar nebula.

A mass-balance equation is written for each element, with the concentrations of all gaseous species containing that element appearing as variables. Free-energy data can be used to express each of these concentrations in terms of a product of an equilibrium constant and powers of concentrations of monatomic gaseous component elements, which become the variables when these expressions are substituted into the mass-balance equations. The result is a series of 15 nonlinear equations in 15 unknowns, which are solved at any pressure and temperature by a method of successive approximations. At each pressure, the equation set is solved at successively lower temperatures, and the solutions are tested to see if any one of a list of over 100 crystalline phases has reached saturation. When such a mineral is found, it is assumed to condense at its equilibrium condensation point and to remain in complete chemical equilibrium with the vapor below that temperature. Terms for the new solid phase are inserted into the mass-balance equations, adding another variable, and an additional equation comes from the condition of gas−solid equilibrium.

Obvious limitations to this approach include the assumption of chemical equilibrium, errors in the elemental abundances, the lack of thermal data for possible important species, and the uncertainties in the free-energy data themselves. The errors in the thermal data lead to uncertainties on the order of ± 10 to 15° in the condensation temperatures of most of the phases considered here; such errors could assume major importance if they lead to a change in the order of condensation. Because the work of Grossman (1972) is based on Cameron's (1968) abundance table, and the data for Ca- and Al-bearing phases presented here are based on Cameron's (1973) table, the dependence of condensation temperature on abundances can be estimated by comparison. The year-to-year fluctuations in solar-system abundance data appear to generate uncertainties of ± 15 to 20° in condensation temperatures.

3. RESULTS

Figure 1 shows the calculated distribution of Al between crystalline phases and vapor at a total pressure of 10^{-3} atm, based on Cameron's (1973) abundances. Corundum is the first condensate of the major elements, appearing at 1742 K. Ninety-eight percent of the Al is condensed when melilite forms by the reaction of corundum with the vapor at 1608 K. The fraction of the total Al in melilite continually increases with decreasing temperature as corundum is consumed. At 1533 K, all the remaining corundum reacts with the vapor to form spinel. Below ~1500 K, the fraction of the Al in melilite decreases with falling temperature. This occurs because melilite is a solid-solution series between $Ca_2Al_2SiO_7$ (gehlenite) and $Ca_2MgSi_2O_7$ (akermanite). The first-appearing melilite is nearly pure gehlenite but, as the temperature falls, more and more akermanite is stable in solution in the melilite. The mole percent of akermanite given by the ideal-solution theory is marked along the melilite curve in Figure 1. It is apparent that significant substitution of Mg for Al in the melilite crystal structure begins below ~1500 K, with the displaced Al forming coexisting spinel. The ultimate akermanite content of the melilite may be as high as 100 mole %. Figures 1, 2, and 3 show that the sequence of condensation and reaction is identical from total pressures of 10^{-3} to 10^{-5} atm. With decreasing total pressure, the entire condensation sequence shifts downward in temperature by about 75° per order-of-magnitude pressure change.

The distribution of Ca between crystals and vapor at 10^{-3} atm is shown in Figure 4. Perovskite is the first Ca-containing condensate, appearing at 1632 K. Although its formation consumes virtually all the Ti in the system by 1550 K, it cannot account for very much of the total Ca because the Ti/Ca ratio in the system is very low. The major sink for Ca at high temperatures is melilite, with over 90% of the Ca condensed at 1550 K. At equilibrium, melilite should react completely at 1442 K to form $CaMgSi_2O_6$ (diopside), although there is some evidence in meteorites that this pyroxene mineral may contain large quantities of Al and Ti in solid solution.

At the time of writing, lower temperature calculations with Cameron's (1973) abundances had not yet been performed; but, in order to examine the condensation of Mg, Si, and Fe, the data of Grossman (1972) can be used, if we bear in mind that these are based on Cameron's (1968) abundance data, which yield slightly higher (by 10 to 20°) condensation temperatures owing to a lower hydrogen abundance.

Figure 5, taken from Grossman (1972), shows that a large fraction of the total Mg begins to condense only when forsterite appears at 1444 K. This phase later reacts with the vapor to form enstatite at 1349 K. Also, metallic iron does not begin to condense until 1473 K. Figure 6 shows that very little of the Si in the system is condensed until forsterite appears. A temperature range of more than 75° exists, from 1550 to 1473 K, over which Ca, Al, and Ti are virtually totally condensed while none of the Fe and less than 15% of the Mg and Si have condensed. There is thus a stage in the cooling history of the solar system in which the condensate is dramatically enriched in Ca, Al, and Ti and depleted in Mg, Si, and Fe, compared to the solar abundances. If the physical conditions in the nebula allowed the efficient separation of dust from gas at high temperatures, some regions of the nebula might be expected to be enriched in Ca, Al, and Ti relative to Mg, Si, and Fe.

4. INTERPRETATION

4.1 Ca−Al-Rich Inclusions in Allende and Other Carbonaceous Chondrites

Although several reports of white, Ca−Al-rich inclusions in carbonaceous chondrites appeared in the meteorite literature before 1969, relatively little significance was attached to these aggregates until the fall of the Allende meteorite in northern Mexico in February of that year. Several tons of this meteorite are estimated to have fallen (Clarke et al., 1970), and it contains between 5 and 10% white inclusions. Figure 7 shows a typical slab surface of Allende. Note the abundance of irregularly shaped, light-colored inclusions, many of which tend to be larger than the more common, darker colored, spherical ferromagnesian chondrules.

4.1.1 Chemistry and mineralogy

In Table 1, the bulk chemical compositions of two Allende inclusions are compared to the calculated equilibrium condensate compositions at several temperatures at 10^{-3}-atm total pressure. The measured compositions, being very rich in Ca, Al, and Ti and poor in Mg, Si, and Fe, are quite close to possible condensate compositions in the temperature range 1440 to 1475 K. There is strong evidence that the small amounts of Na_2O and FeO in some inclusions, such as those in Table 1, are due to reactions between the inclusions and adjacent mineral phases within the meteorite parent bodies long after condensation was over. In addition to the striking similarity in chemical composition, the predominant mineral assemblage of the inclusions is melilite (usually containing less than 35 mole % akermanite) + spinel + perovskite + diopside. These minerals are all predicted to condense from the solar nebula in the temperature range 1632 to 1442 K at 10^{-3} atm.

4.1.2 Textures

Figure 8 is a photomicrograph of a polished surface of an Allende inclusion. In it, gehlenite and spinel are intimately intergrown, tiny perovskite crystals are found inside spinel grains, large cavities are present, and diopside coats the walls of the cavities. Kurat (1970) described similar inclusions containing tiny grains of Al_2O_3 in the Lancé carbonaceous chondrite. He also found diopside rims surrounding entire inclusions. The textural features can be understood in terms of a condensation origin for the inclusions. Figure 9 is a pictorial representation of one possible sequence of events that could have led to the formation of the observed textures during condensation. At 1700 K, corundum would have been the only condensate of the major elements. At 1610 K, it would have been joined by perovskite crystals. The corundum would have begun to react to form gehlenite by the time the temperature fell to 1590 K, and this reaction would have gone nearly to completion by 1550 K. The interior cavities that are now observed may have had their origin where large, irregular grains came together. At 1510 K, the first sign of disequilibrium is shown. Corundum should have reacted completely with the vapor to form spinel at 1533 K, but this reaction may have been impeded by the inability of Mg to diffuse to the corundum in the interiors of the inclusions. Between 1510 and 1440 K, chemical equilibrium demands that the melilite increase its Mg content dramatically. This must take place by the impingement of gaseous Mg and Si atoms on the surfaces of the melilite grains, their diffusion into the grain interiors, their displacement of Al from the melilite crystal structure, the diffusion of Al outward, and finally, the precipitation of Al as coexisting spinel. We would expect to find, as observed, an intergrowth of spinel and melilite and, if the cooling rate were rapid relative to diffusion rates, melilite containing much less than the maximum allowable Mg content. Finally, at 1442 K, melilite should have reacted completely with the gas to form diopside. Thus, the exposed outer surfaces of the inclusions as well as the walls of the interior cavities may have been coated with diopside, but the reaction might not have been able to proceed into the centers of the inclusions. Thus, the textural features observed in the Allende inclusions could have been produced during condensation. Many inclusions, however, do not show the textural relations described here. This may be because they melted and recrystallized after condensation.

Figure 10 is a scanning electron photomicrograph of a Ca-rich inclusion in the Murchison chondrite. It shows that the walls of the cavities are lined with perfect, euhedral crystals, suggesting that they grew from a fluid phase that once filled the spaces. Thus, the cavities are not artificially produced during the cutting of the meteorite.

4.1.3 Trace elements

Even though they would have very low partial pressures in a gas of solar composition, some trace elements are so refractory or form such refractory compounds that they should have condensed above or in the same temperature range as the highest temperature condensates of the major elements. Table 2 lists some of the refractory trace elements and the Ca—Al—Ti-rich condensates in order of decreasing condensation temperature. The condensation temperatures of the major phases are as calculated here from Cameron's (1973) abundances, but the data for the trace elements come from Grossman (1973), based on the calculations of Grossman (1972) and the abundances of Cameron (1968). The temperatures given for the trace elements are probably within 20 to 30° of values consistent with the Cameron (1973) abundances. These elements span a wide range of chemical properties and behavior. For example, in meteorites, W, Re, and the platinum-group metals (Os, Ir, Ru) are nearly always found heavily concentrated in metallic nickel-iron because of their marked chemical affinity for this phase. Similarly, Mo is usually found associated with sulfide minerals, while Zr, Hf, Y, Sc, and the rare-earth elements favor the silicate phases of meteorites. In the final column of Table 2, however, it is clear that _all_ these elements are enriched in the metal—sulfide-free Allende inclusions by factors of 10 to 24 relative to their mean concentrations in Type I carbonaceous chondrites. The only property common to them all is that they have very high condensation temperatures in a gas of solar composition. Perhaps grains of the refractory trace elements acted as condensation nuclei for the Ca—Al—Ti-rich phases, or perhaps the latter acted as nuclei for some of the trace elements. In some cases, such as the rare earths, condensation may have taken place by solid solution of a trace element in the crystal structure of a major phase. These mechanisms could have provided the opportunity for the observed association of the refractory trace elements with the refractory major elements, thus lending additional support for the proposed condensation origin of the white inclusions.

4.2 Fractionation of Refractory Elements

4.2.1 Chondrites

Ca, Al, and Ti, as well as several of the trace elements in Table 2, are known to be depleted in the ordinary and enstatite chondrites by mean factors of 0.69 and 0.50, respectively, relative to the carbonaceous chondrites. Larimer and Anders (1970) suggested that this could have been caused by removal of refractory condensates in the form of white inclusions from those regions of the nebula where the ordinary and enstatite chondrites accreted. Grossman (1973) pointed out that 23 and 38% loss of high-temperature condensates is necessary to account for the depletion factors in the ordinary and enstatite chondrites, respectively. Apparently there did exist physical processes capable of transporting Ca−Al-rich grains away from their condensation site in the nebula before Mg, Si, and Fe condensed.

4.2.2 The moon

Most lunar surface rocks returned by the Apollo missions are highly enriched in refractories and depleted in volatiles (Ganapathy et al., 1970; Wänke et al., 1973). This has led several workers to conclude that the entire moon is enriched in high-temperature condensates, similar to the Allende inclusions, compared to the solar-system abundances (Gast and McConnell, 1972; Anderson, 1972, 1973; Wänke et al., 1973), although some features of lunar geochemistry are inconsistent with this hypothesis (Grossman and Larimer, 1974). This possibility, however, would again suggest that refractory grains were able to accumulate in certain parts of the solar nebula before the condensation of Mg, Si, and Fe.

4.2.3 Interstellar gas

Herbig (1970) and Field (1974, this volume) have presented evidence that some interstellar gas clouds are depleted in such refractory elements as Ca, Ti, and Al by factors of several hundred to several thousand, while the concentrations of more volatile elements, such as the alkali metals and sulfur, are within a factor of 5 of normal (i.e., solar) abundance. They have suggested that the material in these clouds has been processed through relatively dense stellar nebulae and returned to the inter-stellar medium as a mixture of gas and dust. The dust is composed of the refractory elements, which condensed inside the nebulae and are no longer spectroscopically observable, while the volatile elements are still gaseous. The abundances in the inter-stellar gas suggest that, when material is ejected from nebulae, the very refractory elements have almost always condensed, Mg, Si, and Fe have sometimes condensed, and the very volatile elements, such as the alkalis, have only rarely condensed. Thus, many of the grains in the interstellar medium may be very similar to those described here from the Allende meteorite.

4.2.4 Circumsolar condensation today

Hemenway (1974, this volume) has performed qualitative analyses of submicron-sized dust grains collected at high altitudes. These particles are believed to be solar in origin and are very rich in rare earths, Hf, W, Th, Ta, Ca, and Ti. From Table 2, these elements are among the highest temperature condensates from a gas of solar composition. It seems quite possible that circumsolar condensation may still be taking place today in relatively cool, dense regions of the solar atmosphere and that the observed grains are pushed away from their condensation site by radiation pressure. The identification of the refractory platinum metals in these grains would provide further evidence of their condensation origin.

4.3 Interstellar Grains in Meteorites

Onuma, Clayton, and Mayeda (1972) presented a model for the variation with temperature of the O^{18}/O^{16} ratios of various mineral phases in chemical equilibrium with a cooling gas of solar composition. The model is based on the fact that different isotopic species of the same molecule have slightly different thermodynamic properties, and therefore, isotopes of the same element fractionate from one another very slightly during normal chemical reactions.

In the notation of oxygen-isotope geochemistry,

$$\delta O^{18}_{samp} = \left[\frac{(O^{18}/O^{16})_{samp} - (O^{18}/O^{16})_{std}}{(O^{18}/O^{16})_{std}} \right] \times 10^3 \ ,$$

where samp indicates a sample and std refers to a standard, which is usually SMOW (standard mean ocean water). Onuma, Clayton, and Mayeda (1972) suggested that, as grains of forsterite, enstatite, or melilite equilibrated their oxygen isotopic compositions with the cooling solar nebular gas, $\delta O^{18}_{mineral} - \delta O^{18}_{gas}$ should have decreased at high temperature, reached a minimum value of -9 to $-12\,\%_O$ at about 800 K at 10^{-3} atm, and then increased again at lower temperature.

It can be shown from statistical mechanics that, as a result of normal chemical equilibrium isotope fractionations, such as might have taken place during condensation, a graph of δO^{17} vs. δO^{18} should be very nearly a straight line with a slope of $+1/2$ because the isotope effects are almost linearly proportional to the relative mass difference of the isotopes, $\Delta m/m$. This is seen to be the case in Figure 11 for lunar and terrestrial rock samples, ordinary chondrites, and achondritic meteorites [line (a)]. This implies that the oxygen isotopic compositions of these samples are related, perhaps in a very complex way, through a series of chemical reactions in which oxygen isotopic equilibration was achieved.

Clayton, Grossman, and Mayeda (1973) showed that many of the suspected high-temperature condensate minerals (melilite and forsterite) in carbonaceous chondrites have δO^{18} substantially below the minimum value attainable in the Onuma, Clayton, and Mayeda (1972) condensation model. These phases, labeled C2 and C3 anhydrous minerals in Figure 11, plot along a line of slope $+1$ [line (b)], indicating that their isotopic compositions are <u>not</u> related by chemical isotope effects. The large depletions of O^{18} and O^{17} relative to O^{16}, compared to lunar and terrestrial rocks, are thus due to the intervention of nuclear, rather than chemical, effects. The consumption of O^{18} and O^{17} relative to O^{16} in nuclear reactions, caused by cosmic-ray bombardment or occurring in some hypothetical early stage of the sun, seems precluded by the slope of $+1$, which requires equal fractional depletions of both O^{18} and O^{17} and, in turn, identically equal cross sections for the two postulated nuclear reactions. The simplest interpretation of the oxygen isotope ratios of the high-temperature minerals is that line (b) is a mixing line, with each inclusion containing two distinctly different components of its oxygen. The major component has a "normal" solar-system oxygen isotopic composition at the intersection of lines (a) and (b). Added in varying but small amounts to each inclusion is a second component, dramatically depleted in O^{18} and O^{17} relative to O^{16}. It is probably pure O^{16} because the slope of $+1$ means that, if O^{18} and O^{17} were present in the second component, its O^{18}/O^{17} ratio would have to be the same as the solar-system ratio, an unlikely coincidence.

Suppose a presolar-system primitive star existed that contained only one isotope of oxygen, O^{16}. If circumstellar, oxygen-containing condensate grains were later ejected by this star into an interstellar region, they could have mixed with the gaseous and crystalline debris of other stars having different isotopic compositions. If the solar system formed by the gravitational collapse of this cloud, and if peak temperatures in the solar nebula were never high enough or long enough to evaporate these grains completely, they or their evaporation residues could have survived the birth of the solar system, preserving a clue to their unique nucleosynthetic history in the form of an isotopic composition that was never homogenized with the rest of the material in the vapor of the solar nebula. Mixing line (b) may then have been generated when high-temperature minerals condensed from the gas, nucleating on the preexisting interstellar grains.

This is the strongest evidence to date of the presence of interstellar grains in meteorites. Isotopic anomalies of elements other than oxygen in the same samples are expected and should provide clues to the nucleosynthetic processes that occurred in the interiors of primitive stars in the Galaxy. A complete list of elements that show such anomalies can yield clues to the mineralogy of the surviving interstellar grains and, therefore, to the peak temperatures reached during the contraction of the solar nebula.

CONCLUSIONS

The major- and trace-element chemical composition, mineralogy, and textural features of the white, Ca-rich inclusions in Allende and other carbonaceous chondrites strongly support the contention that they are the highest temperature condensates from the solar nebula. Because the sun is similar in composition to many other stars in the Galaxy, because interstellar grains are thought to have formed inside stellar nebulae, and because the interstellar vapor phase is depleted in the refractory elements, the chemical and physical properties of the meteoritic phases may be very similar to those of interstellar dust. The systematics of the oxygen isotope variations of the solar nebula condensates in carbonaceous chondrites suggest the direct incorporation of interstellar grains in the chondritic mineral phases. It is evident that there is much to be gained from the interchange of ideas between astronomers and cosmochemists.

ACKNOWLEDGMENTS

The writer is grateful to Dr. A. V. Crewe for the use of his scanning electron microscope facilities and to Mr. T. J. Peterson, electronics engineer, and Dr. S. D. Lin for their technical assistance. Mr. Frederic Oldfield helped with the computer calculations.

This work was partially supported by funds from the Research Corporation, the Louis Block Fund of the University of Chicago, and the National Aeronautics and Space Administration through grant NGR 14-001-249.

REFERENCES

Anderson, D. L., 1972. Nature 239, 263.

Anderson, D. L., 1973. Earth Planet. Sci. Lett. 18, 301.

Cameron, A. G. W., 1968. In Origin and Distribution of the Elements, ed. by L. H. Ahrens (Pergamon, New York), p. 125.

Cameron, A. G. W., 1973. Space Sci. Rev. 15, 121.

Cameron, A. G. W., and Pine, M. R., 1973. Icarus 18, 377.

Clarke, R. S., Jr., Jarosewich, E., Mason, B., Nelen, J., Gómez, M., and Hyde, J. R., 1970. Smithsonian Contr. Earth Sci. No. 5, 53 pp.

Clayton, R. N., Grossman, L., and Mayeda, T. K., 1973. Science 182, 485.

Fuchs, L. H., 1971. Amer. Mineral. 56, 2053.

Ganapathy, R., Keays, R. R., Laul, J. C., and Anders, E., 1970. Geochim. Cosmochim. Acta, Suppl. 1, 1117.

Gast, P. W., and McConnell, R. K., Jr., 1972. In Third Lunar Sci. Conf. Abstracts (Lunar Science Institute, Houston), p. 257.

Grossman, L., 1972. Geochim. Cosmochim. Acta 36, 597.

Grossman, L., 1973. Geochim. Cosmochim. Acta 37, 1119.

Grossman, L., and Larimer, J. W., 1974. Rev. Geophys. Space Phys., in press.

Herbig, G. H., 1970. Mem. Soc. Roy. Sci. Liège, Ser. 8, 19, 13.

Kurat, G., 1970. Earth Planet. Sci. Lett. 9, 225.

Larimer, J. W., and Anders, E., 1970. Geochim. Cosmochim. Acta 34, 367.

Onuma, N., Clayton, R. N., and Mayeda, T. K., 1972. Geochim. Cosmochim. Acta 36, 169.

Wänke, H., Baddenhausen, H., Dreibus, G., Quijano-Rico, M., Palme, H.,
Spettel, B., and Teschke, F., 1973. In Lunar Science IV, ed. by J. W.
Chamberlain and C. Watkins (Lunar Science Institute, Houston), p. 761.

TABLE 1.

Chemical compositions of Allende white inclusions compared to calculated condensate compositions at $P_{tot} = 10^{-3}$ atm.

	Condensate 1475 K	Coarse-grained inclusion[*]	Condensate 1450 K	Fine-grained inclusion[†]	Condensate 1440 K
CaO	32.31	26.76	27.23	21.6	18.86
Al_2O_3	34.81	31.61	29.22	26.6	20.21
TiO_2	1.77	0.99	1.49	1.3	1.02
MgO	9.39	10.82	16.98	13.1	21.03
SiO_2	21.71	29.79	25.09	33.7	38.87
Total	99.99	99.97[‡]	100.01	96.3[§]	99.99

[*]Clarke et al. (1970), type a chondrule NMNH 3529.

[†]Clarke et al. (1970), single aggregate NMNH 3510.

[‡]Also contains 0.37% FeO and 0.11% Na_2O.

[§]Also contains 0.1% Cr_2O_3, 2.3% FeO, and 1.1% Na_2O.

TABLE 2.

Condensation temperatures of refractory trace elements
compared to those of the major high-temperature minerals.

Crystalline phase	Condensation temperature (°K)		Enrichment factor[*] in Allende inclusions
	10^{-3} atm	10^{-4} atm	
Os	1925	1840	15
W	1885	1798	13
ZrO_2	1840	1789	10
Re	1839	1759	21
Corundum (Al_2O_3)	1742	1671	
HfO_2	1719	1652	10
Y_2O_3	1719	1646	21
Sc_2O_3	1715	1644	22.9[†]
Mo	1684	1603	10
Perovskite ($CaTiO_3$)	1632	1557	
RE_2O_3 (in solution)[‡]	1632	1557	22.5[†][§]
Ir	1629	1555	24.1[†]
Ru	1614	1541	12
Gehlenite ($Ca_2Al_2SiO_7$)	1608	1528	
V_2O_3	1534	1458	
Spinel ($MgAl_2O_4$)	1533	1451	
Ta_2O_5	1499	1452	18
ThO_2	1496	1429	
Diopside ($CaMgSi_2O_6$)	1442	1375	

[*] Enrichment factor = ratio of concentration in Allende inclusion to concentration in Type I carbonaceous chondrites.

[†] Data from Grossman (1973), who analyzed 16 inclusions. All other enrichment factors are based on one Allende inclusion, analyzed by Wänke et al. (1973).

[‡] RE = rare-earth elements. The rare earths can condense in solid solution in perovskite.

[§] Average of La, Sm, Eu, Yb.

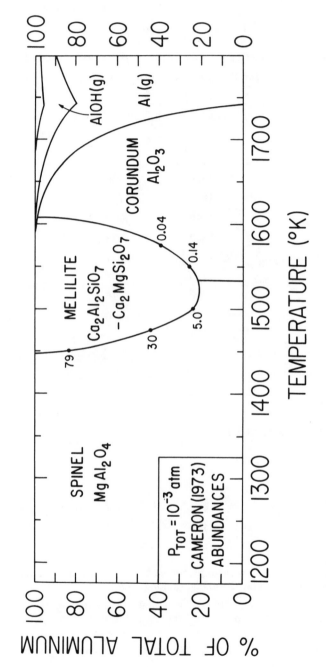

Figure 1. The calculated distribution of Al between crystalline phases and vapor as a function of temperature in a system of solar composition at a total gas pressure of 10^{-3} atm. Over 90% of the Al is condensed at 1650 K. Numbers on the melilite curve are the concentrations of $Ca_2MgSi_2O_7$ in the melilite, in mole percent.

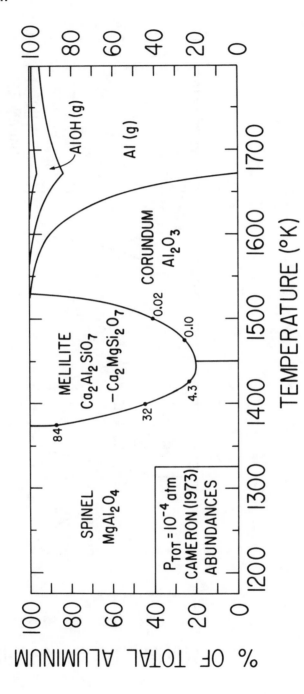

Figure 2. The calculated distribution of Al at 10^{-4} atm. The sequence of condensation and reaction is the same as in Figure 1 at 10^{-3} atm, but is shifted downward in temperature by 70 to 80°.

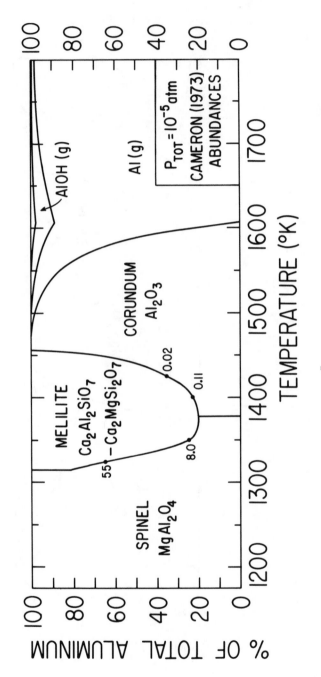

Figure 3.   The calculated distribution of Al at 10^{-5} atm.   The sequence of condensation and reaction is the same as in Figure 2 at 10^{-4} atm, but is shifted downward in temperature by 65 to 75°.

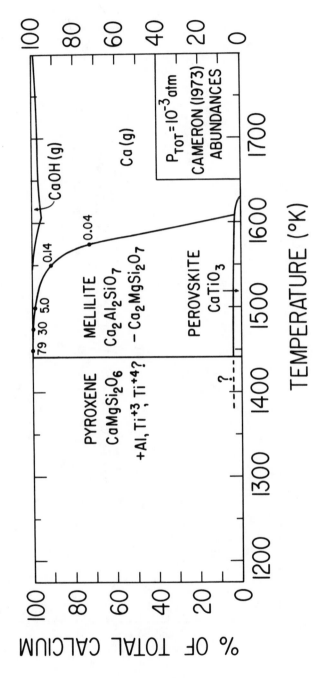

Figure 4. The calculated distribution of Ca between crystalline phases and vapor as a function of temperature in a system of solar composition at a total gas pressure of 10^{-3} atm. Ninety percent of the Ca is condensed at 1550 K and 90% of the Ti is condensed at 1600 K.

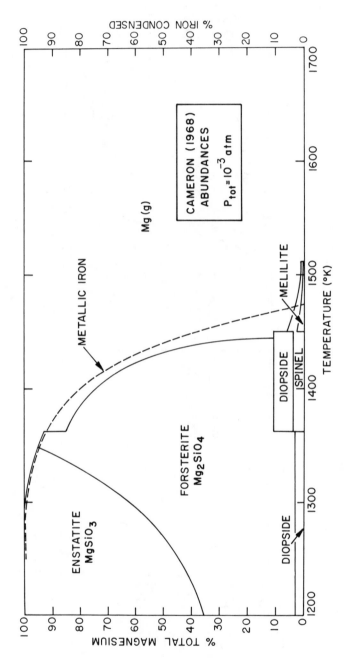

Figure 5.   The calculated distribution of Mg between crystalline phases and vapor as a function of temperature in a system of solar composition at a total gas pressure of 10⁻³ atm. Although this figure is based on older abundance estimates and thermodynamic data than are Figures 1 through 4, the temperatures shown here are only about 10 to 20° higher than would be consistent with these earlier figures. The metallic iron condensation curve is shown for reference. Significant condensation of iron begins to occur below 1470 K and that of magnesium, below 1440 K. From Grossman (1972).

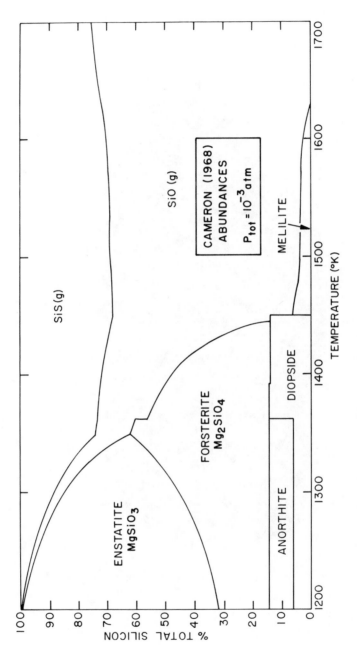

Figure 6.   The calculated distribution of Si between crystalline phases and vapor as a function of temperature in a system of solar composition at a total gas pressure of 10^{-3} atm.   This figure is based on the same data as is Figure 5.   Significant condensation of Si begins below 1440 K.   From Grossman (1972).

Figure 7. Slab surface of the Allende meteorite. Many of the light-colored inclusions visible in this photograph are the Ca–Al-rich ones discussed in the text. Note the complex variety of different inclusions in terms of size, shape, and color. The Ca–Al-rich inclusions represent only between 5 and 10% of the meteorite by volume. Photo courtesy of R. Ganapathy.

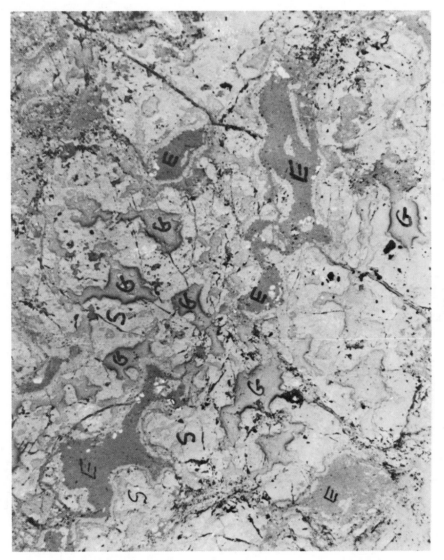

Figure 8. Photomicrograph of a polished surface of a Ca−Al−rich inclusion in the Allende meteorite. Areas marked G are gehlenite ($Ca_2Al_2SiO_7$) and S are spinel ($MgAl_2O_4$). White specks in spinel are perovskite ($CaTiO_3$). Epoxy-resin-filled cavities (E) are lined with dark rims of diopside ($CaMgSi_2O_6$). Width of field is 1.0 mm. Photo from Fuchs (1971), courtesy of the author.

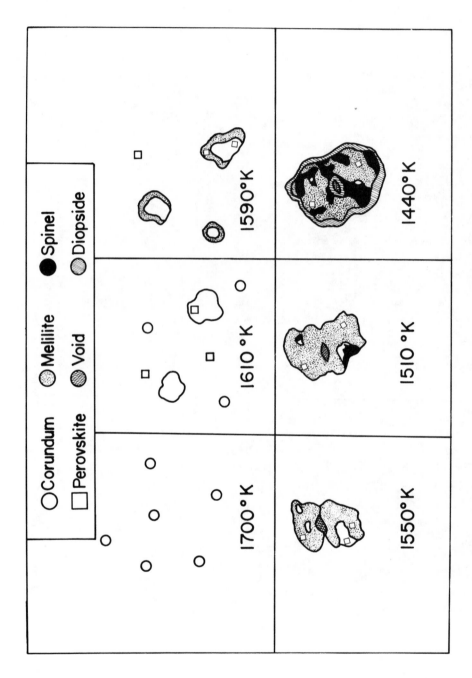

Figure 9.   Possible sequence of condensation reactions leading to the textures observed in the Ca—Al-rich inclusions in Allende, such as the one shown in Figure 8.   Temperatures refer to a total pressure of 10^{-3} atm.

Figure 10. Scanning electron photomicrograph of a Ca–Al-rich inclusion in the Murchison chondrite. The euhedral crystals lining the walls of the cavity indicate crystal growth from a fluid that once filled the cavity. Width of field is 4.4 μ.

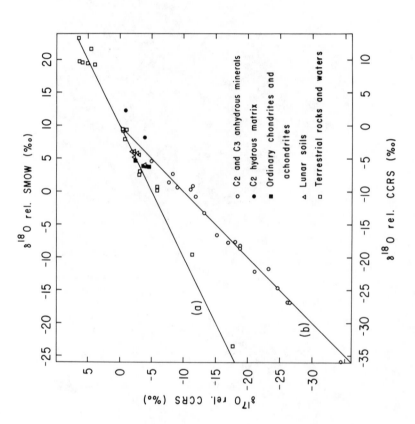

Figure 11. Oxygen isotopic compositions of various meteoritic, lunar, and terrestrial materials. Ordinary chondrites, achondrites, and terrestrial and lunar rocks are plotted along line (a) with a slope of + 1/2, indicating that their isotopic compositions are related to one another through a series of chemical processes. The C2 and C3 anhydrous minerals are the high-temperature condensates referred to and are plotted along line (b) with a slope of +1. These specimens are mixtures of "normal" solar-system oxygen having an isotopic composition at the intersection of the two lines and an interstellar O^{16}-rich component. From Clayton, Grossman, and Mayeda (1973).

COMMENTS

Fred L. Whipple

It is indeed a great honor for me that so many of you have come together to discuss a subject that has been dear to my heart for some four decades. I am especially grateful to George Field and to Al Cameron for the concept of this symposium and for their initiative and support in bringing it about.

Time permits me to discuss only a few aspects of the many topics that have been opened up during these past three days. I will begin with the problem of chondrules, first introduced by Al Cameron. The problem is to find a mechanism for producing quickly cooled and supercooled droplets in a solar nebula and, particularly, for producing them in such huge quantities. Some chondrites actually contain 70 to 80% chondrules by mass. It is not really true that I have given up the idea of making chondrules by lightning (Whipple, 1966), that is, by electrical discharge, although I have never been greatly enamored of the idea. It always struck me as one of those "desperation" solutions that the scientific detective makes after all the more rational solutions to the problem seem to have failed. In the solar nebula, at a temperature of some 500 K, it is quite impossible to establish the large electrical potential gradients needed to produce electrical discharge. Enough electrons will be produced to "bleed" off the electrical potential because of the low densities involved. On the other hand, when we consider small asteroidal bodies moving at finite velocities through the nebula, we find a drastically different situation.

Suppose that dust in such a nebula carries charges that are correlated with the dimensions of the particles. The bow wave or shock wave, depending on the velocity of the asteroid, will separate charge as it separates the dust according to size. Large particles will be accelerated more slowly than small particles in the gas of the bow wave. Thus, it is possible in the bow wave to set up rather large electrical potential gradients on a short time scale, of the order of seconds. The physical situation is not too different from that in atmospheric clouds, where circulation currents produce the same result on water droplets that carry charge. Because of the short time constant, the few electrons present are not able to drain off the potential difference, as one would expect, over large distances and long periods of time in the normal circulation or turbulence of the solar nebula.

Making chondrules in the bow wave of an asteroidal body moving through the solar nebula places the chondrules in precisely the most advantageous region of space for their capture on the asteroid in the accretion process. As I have shown, such a moving asteroid can aerodynamically capture solid particles above a certain critical size, depending on the velocity of motion and the density of the nebula (Whipple, 1972). The finer dust is swept away by the flow pattern around the asteroid. Thus, if the dust in the nebula is rather finely divided, it might frequently happen that the major component of accretion would be the chondrules made in the bow wave by electrical discharge. No theoretical attention has been given to this possible process of electrical discharge. It deserves further investigation, as do related aerodynamic processes that will affect the production and capture of chondrules by impact splashing. Ed Purcell has discussed problems concerning charges on dust particles in space. I believe his thinking might be of extreme importance in the theoretical study of potential difference and possible electrical discharge in the bow-wave problem.

The concept of chondrule production by the splashing of impacting bodies was mentioned by Harold Urey. If we assume that the process occurred at a time when the ambient temperature was moderately high, perhaps somewhat greater than 500 K, then the process could be relatively efficient, and the cooling time for the droplets would be adequate for them to assume the characteristics that we observe in chondrules. Even so, I have difficulty visualizing a situation in which chondrules made by impact splashing and accumulated on the body could contribute 70 to 80% of the mass accreted. Furthermore, as John Wood points out here, there are chondrule-like droplets on the moon but the efficiency of production is extremely low, of the order of one chondrule for 100 bonded agglomerates. Thus, the resultant breccia does not much resemble chondritic meteorites. Impact splashes on a relatively hot solid near the melting point might, however, produce quite a different result.

Winfield Salisbury has been experimenting in the laboratory with an electrical discharge in an atmosphere containing dust particles. He finds that the <u>pinch effect</u> does indeed make chondrule-like glassy spherules with an efficiency of the order of 1%. His experiments have been carried out with granite dust and with a mixture of the earthy elements approximating that of the solar abundances. It has yet to be proved that internal structure of these spherules is really like that of chondrules and that the crystalline and glassy characteristics are similar. Ursula Marvin reports in the discussion that in the few studies she and John Wood have made, the composition was not like that of chondrules. It is quite likely that studies made so far involve the granite source of particles rather than the solar mix.

In answer to a question by George Field, heating in the shock wave of an asteroid moving through the solar nebula would not, in my opinion, be effective in producing chondrules. It is too difficult to heat the shock wave sufficiently and for long enough that both melting and aggregation of dust could take place or that larger initial bodies could melt into glassy spherules. Also, it would heat up the matrix material captured at the same time. In chondrites, the matrix has not been heated so much as the chondrules. I am not aware, however, of any careful calculations that show these processes to be impossible, especially under extreme conditions of nebula density, velocity, and body size.

Field's question suggests to me another physical mechanism that should be explored. A dust particle in the solar nebula impinging on the bow wave of a moving asteroidal body simulates a slow meteor in the earth's atmosphere. Perhaps at a moderately elevated ambient temperature augmented by the pressure heating in the bow wave, the aerodynamic friction on the impinging dust particle might heat it enough to melt it into a chondrule. The heating and the meteoric and aerodynamic effects under given physical conditions would be highly dependent on particle size.

In concluding this discussion of chondrules, I might say that we need more theoretical work, particularly with regard to 1) charge separation and electrical discharge in the bow wave of a moving asteroid, 2) heating by a supersonic shock wave, 3) the meteoric heating mentioned above, and 4) the aerodynamics of impact splashing on bodies at modestly elevated temperatures in an ambient gas at similar temperatures still below the melting point of chondritic material. On the laboratory side, it is clear that Salisbury's spherules produced by electrical discharge should be studied more thoroughly to ascertain whether they are indeed like chondrules in structure and whether modifications of the discharge or the ambient gas pressure might be more favorable to the production of truly chondrule-like spherules.

I should now like to bring up a subject not discussed in the conference, but particularly relevant to the work that Purcell has reported. In meteor streams, where we have good information about the time scale of cometary deposition of meteoritic material in space, all the evidence indicates that the subsequent spatial dispersion is higher than we calculate. The spreading of the stream is theoretically based on the known physical processes with which we are acquainted — such as collisions at high velocities, the Poynting-Robertson effect, sputtering by the solar wind, and light-pressure effects. It looks as though some unknown processes disperse meteor streams.

Both Richard Southworth (1963) and Salah Hamid (1950) wrote theses on this subject for the Perseid meteor stream. I feel that the processes are those that Ernst Öpik (1951) has talked about, the effect on small particles of high spinning rates coupled with differential radiation effects from different areas on the particles or, possibly, collisional effects arising from the rotation that will in some way disperse bodies in meteor streams. The great Taurid stream appears to be another example in which the dispersive effects exceed our theoretical expectations (Whipple and Hamid, 1951).

George Field yesterday asked me to define the icy-comet model. When I published the concept (Whipple, 1950), I did not feel it was such a novel idea, but apparently it had never been enunciated previously. I often wondered why this was true and finally realized that few other people were working in fields that would make the necessity for a discrete icy nucleus so demanding. In the late 1940s, we had no clear concept of the density of the interplanetary medium, and it always seemed possible that the particles in a gravel-bank nuclear model of a comet might reabsorb gas from the interplanetary regions during the long periods spent near aphelion. Also, we believed that the greater fraction of the mass in a comet was carried by the meteoritic particles.
J. L. Gossner and I showed that, from the polarization of the zodiacal light, the density

of electrons in near regions of space could not exceed 1000 cm^{-3} (Gossner and Whipple, 1949). Although this limit is extremely high compared to the modern value of ~5 cm^{-3}, it nevertheless provided gravel grains with no means whatsoever for accumulating gases to replace those lost via deadsorption by solar heating near perihelion. There clearly had to be a reservoir of material to provide the activity of periodic comets over many revolutions. Halley's Comet dates back at least 29 revolutions to 239 B.C. (Kiang, 1972), and Encke's Comet had clearly made at least 1000 revolutions without completely losing its activity (Whipple and Hamid, 1951).

Next, the persistence of sun-grazing comets proved the icy-reservoir concept. It is quite impossible for a gravel bank to survive a close approach to the sun at one-third the solar diameter. The solar heat would vaporize the gravel particles, never to be reconsituted. Furthermore, tidal disruption within the Roche limit would complete the destruction of sun-grazing comets. It is known that a gravel bank is gravitationally a fluid. Nevertheless, a number of these comets with a period of somewhat less than 1000 years have returned.

It was not these rather obvious considerations that led me to publish the icy-conglomerate model, but the fact that it provided an explanation for the nongravitational motions of comets. Comet Encke, in particular, exhibited a shortening of its 3.3-year period by some 2.5 hours per period during the early part of the 19th century. Frances Wright and Whipple (1950) had associated Comet Encke with the huge Taurid meteor stream, which indicated a very long lifetime for the comet and an enormous contribution of meteoritic material to the interplanetary medium (Whipple and Hamid, 1951). As an icy-comet nucleus with imbedded dust approaches the sun, the solar heat causes it to sublimate away the ice, which carries off the dust. This produces a jet action, which, in a rotating comet, can produce a force perpendicular to the radius vector toward the sun. Whether the motion of the comet will be retarded or accelerated will depend on the sense of the nuclear rotation. I assumed a reasonable efficiency in the heating process, with a consequent loss of much more material than was observed either as dust or gas in periodic comets. By losing a small fraction of 1% of the mass per period, a faint comet like Encke can produce the necessary force to account for the nongravitational motions.

Hamid and Whipple (1953) also showed that the expected major force of jet action on the comet nucleus radially away from the sun is statistically evident in the motions of well-observed comets with longer periods. These results have been confirmed abundantly by Brian Marsden (1968, 1969, 1970). Marsden and Sekanina (1971, 1974) have also shown that the secular change in period for Comet Encke has markedly decreased during the past century and a half, and that almost all well-observed short-period comets show nongravitational effects, half acceleration and half retardation, in their motions. The proof of the large loss of mass has been finally provided by observations in space, specifically those by Code (Code, Houck, and Lillie, 1972) and by Bertaux and Blamont (1970). These observations, made in early 1970, demonstrate the presence of a huge dissipating hydrogen cloud and, probably, a comparable number of OH molecules about these comets; these findings are consistent with the predictions by Ludwig Biermann (Biermann and Trefftz, 1964) on the basis of water-ice sublimation and confirm the basic postulates of the icy-comet model.

The detailed structure of an icy-comet nucleus was vague to me then and remains vague, because we still do not have enough information to solve the problem. Boris Levin criticized me lightly and justly for simultaneously presenting three concepts of the structure. In one of these, the dust is imbedded as pieces in the icy matrix. In another, the nucleus is all matrix, consisting of loosely connected earthy material interspersed with icy material. In the third, the material is more clumpy with ice and loose aggregates of meteoritic material. The Harvard photographic meteor studies by Richard McCrosky (1955, 1958), Luigi Jacchia (1955), and Allan Cook (Cook, Jacchia, and McCrosky, 1963; Whipple, 1962) demonstrated that the meteoroids producing the ordinary meteor streams and sporadic meteors are low-density fragile bodies. Peter Millman (1974, this volume) shows that, for four of the most common elements, the relative abundances are not distinguishable from a solar mix. Thus, it is abundantly clear that the "solid" particles in comets are not well-condensed meteoritic material but very loose low-density aggregates of weak structure. This has been borne out by the friable character of the bodies producing the brilliant fireballs in the Prairie Network program conducted by McCrosky (McCrosky et al., 1971).

The early evidence that I produced (Whipple, 1955) to demonstrate that comets contribute the major part of the meteoritic material to maintain the interplanetary complex has been verified by more recent deep-space experiments. This conclusion is important because the most abundant particles by mass in the distribution of the interplanetary dust are of mass about 10^{-5} g (Whipple, 1967), a fact to be explained. There may also be a considerable contribution at the micron or submicron level, but the lifetime is exceedingly short. These observations bear heavily on the serious problem of the origin of the comets.

A number of us, including Gerard Kuiper, Boris Levin, and Ernst Öpik, have been uneasily satisfied that the comets we now observe were formed originally in the Uranus–Neptune region of the solar system and dispersed into the Öpik-Oort cloud (Öpik, 1932; Oort, 1950) by the perturbations of the giant planets. I am completely satisfied that the mean densities of Uranus and Neptune support the concept that they are, indeed, an aggregate of comets that froze out of the solar nebula. Low temperatures must have existed in that region of space when the planets were forming. If there is a residual comet belt beyond Neptune, as I once suggested (Whipple, 1964), its mass does not exceed 1 earth mass to 50 a.u., as demonstrated by the lack of perturbations on Halley's Comet according to the calculations by Hamid, Marsden, and Whipple (1968). The explanation may possibly lie in the earlier-mentioned tendency of particles in a primitive solar nebula to spiral in the direction of increasing gas density, i.e., toward Neptune from greater proto-solar distances (Whipple, 1972).

First Öpik (1965) and now Edgar Everhart (1973) find great difficulty in accounting for the Öpik-Oort cloud of comets by perturbations of comets from the outer planetary region. I believe this problem deserves further investigation, including specifically the effect of a loss of mass in the inner proto-solar system by the action of a solar gale, such as described in George Herbig's concepts of T Tau stars.

McCrea (1960) suggested that comets may have formed in fragmented interstellar clouds, "floccules," at great distances from the center of random encounter of these floccules where the sun and planets formed. His statement on comets follows: "Comets

are bodies of mass probably much less than that of the smallest satellites and again
the present theory of the solar system would not be expected to have anything to contribute about them. On the other hand, the general ideas put forward about the conditions required for star formation may have some bearing upon the problem of the
origin of comets. For our suggestion is that, before star formation occurs, the interstellar hydrogen would be largely converted into molecular form. Hence any material
left over after the formation of stars and planets would be in a very different form
from typical interstellar gas. This may explain the puzzling fact that some clusters
apparently show no interstellar matter at all. Consequently, the origin of comets
ought presumably to be sought in the modified interstellar material of a cloud that has
been through the state required for star formation and not in typical interstellar material."

My own specific mention of the possibility (Whipple, 1964) was based on my own
not completely dissimilar ideas (Whipple, 1948). Cameron (1962) independently suggested the idea as follows: "It may be noted that a great deal of nebular gas is shed
beyond the orbit of Neptune. Although no major planet condensed in this region, presumably the comets did so. In such a region, the gas density is very low, and the
gas is not very effective in transporting solid bodies around. Whatever the method of
dissipation of this peripheral gas, either inwards or outwards, it is difficult to escape
the conclusion that there must be a tremendous mass of small solid material on the
outskirts of the solar system."

In any case, the concept that the comets we observe were formed in clouds moving
in the region of the Öpik-Oort comet cloud is very tempting. It places comets originally
in large orbits before the gravitational effects of passing stars.

Detailed studies of comets from both the ground and space and, particularly,
from space probes near the nucleus should resolve the problem. Comets formed
in the outer planetary region should show evidence of quasi-equilibrium temperature
conditions so that the parent molecules for the observed cometary radicals might,
for example, be water, ammonia, and methane. The presence of carbon
monoxide and carbon dioxide as well as C_2 and C_3 are not chemically consistent with
these physical circumstances of origin. Also, we face the fact that "new" comets,

in the Oort-Schmidt sense, must contain sizable quantities of material far more volatile than water in order to produce considerable activity at distances greater than 3 a.u. On the one hand, it is difficult to see how temperatures in any solar nebula or in the region of new star formation could ever be low enough to freeze out methane and carbon monoxide, which would require temperatures on the order of perhaps 20 K. Armand Delsemme and Pol Swings (1952) long ago suggested that the highly volatile materials were not frozen out but captured in the water-snow structure as hydrates or, more technically, clathrates. Until we know the relative abundances of the parent molecules with their precise identifications, we do not have a sound theory for explaining the activity of comets at great distances. Although Delsemme (Delsemme, 1972; Huebner and Weigert, 1966) has demonstrated that these more distant comets eject dust in the form of icy grains, we still need to understand the physics of the escape of such material from the surface of a comet.

If comets are indeed formed in nearby fragmented interstellar clouds, then we should expect two observable consequences: 1) dust-particle nuclei surrounded by icy mantles as fundamental building blocks in comets, and 2) more exotic parent molecules for the observed radicals, including many of those now being discovered in interstellar clouds by radio astronomy. Such an icy-comet nucleus would be expected to disintegrate completely under the action of solar radiation. The individual grains should presumably be small. It is thus difficult to explain the large aggregates we observe as meteoroids, particularly those that produce fireballs. On the other hand, the comet producing the Tungusta fall in 1908, as proposed first by F. J. W. Whipple (1930), was almost certainly a very low-density body of the type we now visualize as condensing out of an interstellar cloud.

In the outer planetary region of the solar nebula, we can visualize a situation in which, with lowering temperature, earthy material first condenses into sizable bodies before the ices freeze. All depends on the relative rate of temperature drop to nebular density. Thus, comets may well contain nuclei that are optically indistinguishable from asteroids; we desperately need to determine whether the Apollo, or earth-crossing, asteroids are typical of those in the main belt of the asteroids or whether they are the type of structure we might expect from the inner nucleus of a comet.

A definitive solution to these problems will not be easy owing to the possibility suggested by Whipple and Stefanik (1966) that radioactive materials in the comet nuclei may have redistributed the more volatile substances and, to some extent, altered the primordial physical and chemical constitution. This focuses on the highly relevant observational fact that some comets split into more than one component in deep space.

Among 13 such comets observed to split, three obviously were the result of tidal disruption, two of them being in the sun-grazing family and the third having made a very close approach to Jupiter, all within the Roche limit for extremely weak icy nuclei (see Öpik, 1966). Incidentally, the phenomena of comet splitting strongly support the concept of a discrete cometary nucleus. In several cases, the observers were unable to distinguish the two components by means of their brightnesses. In other words, the component that persisted and must have been the most massive, was at times fainter than the small, highly dissipative component. This follows naturally if we accept the idea that a discrete cometary nucleus generally tends to contain a greater concentration of volatile materials in its interior than on its exterior, where solar radiation has been active. Any solid body broken into two pieces exposes two cross sections that are identical in area. A comet in the process of splitting should brighten, with both nuclei of comparable activity and brightness. If the smaller component should tend to break up further, it might easily lose gas more rapidly than the larger component under the action of solar radiation because of the newly exposed highly volatile surfaces. These deductions are completely consistent with the observations of some of the sun-grazing comets and also the split periodic comets.

Two of the 13 comets observed to split were periodic comets. In the famous case of Biela's Comet, the splitting occurred at some 6 a.u. from the sun (Marsden and Sekanina, 1971), well away from likely collisions with asteroidal material and also from highly exciting effects of intense solar radiation.

The remaining eight comets include two with fairly high orbital uncertainties, and therefore classed as parabolic, and six that appear to be "new" comets, i.e., entering the inner solar system for the first time. Of the six, two have large perihelion distances, >3 a.u., as established by Marsden and Sekanina (1973), and fall into the group of new comets, which are not affected by nongravitational forces. The existence of comet groups other than the sun-grazing group, as judged by J. G. Porter (1963) and others, suggests again that new or fairly new comets can split.   The reality of these other groups requires deeper study than Öpik's (1971) recent work on the subject.

The obvious, simplest, and possibly correct explanation of new split comets is that they were double or multiple before entering the inner solar system.   Tidal, or possibly differential nongravitational forces, broke their weak gravitational ties.   I have not accepted this explanation in the past, because I did not see how such extended binary comets could survive the large perturbations necessary to remove them from the outer planetary region of the solar nebula.   Furthermore, the density of the nebular gas should have provided enough drag to coalesce such double comets.   If, however, we choose the alternative that new comets were originally formed in great orbits (a $\overset{\sim}{>}$ 10,000 a.u.) from fragmented interstellar clouds, both objections to the double-comet hypothesis vanish.   The new comets would never have been subjected to strong differential gravitational stress nor to a high-density ambient nebula.

As Maurice Dubin has suggested in the discussion, new comets internally heated by natural radioactivity, as Stefanik and I postulated, might decrease their moment of inertia.   The result could be a spinup if a very loose snowball with considerable rotation collapsed by heating to a much smaller volume.   This explanation for physical splitting from rotation appears simpler and more natural than the processes that Stefanik and I considered.

Reduction in the moment of inertia per unit mass could not occur, however, for old periodic comets such as Biela's and Taylor's (1916 I).   Collisions, as postulated by M. Harwit (1967, 1968) and by Marsden and Sekanina (1971), provide a possible explanation, but one that I do not readily accept, because of its small probability and my uncertainty about the physics of the process.   I prefer the explanation that these old comets split because of spinup most probably induced by asymmetrical gas ejection in the normal process of mass loss by solar radiation.

In summary, then, I lean toward the working hypothesis that comets split in three distinct fashions: 1) tidal disruption of the nucleus in near approaches to the sun or great planets, 2) possible tidal separation of double new comets on entering the inner solar system, and 3) spinup induced in the nuclei of old periodic comets (and possibly new comets) by asymmetrical gas ejection. All three processes deserve much more careful study. For new comets, three possibilities still remain: temperature shock, explosive pockets of gas or voltatiles, and collisions. A much more thorough study should be made of the suggestion by Donn and Urey (1956) that free radicals could provide a powerful source of energy in comet nuclei that would cause comet outbursts. This energy might be triggered in certain volumes of a comet, either by external (solar) or internal (radioactive) heating, to produce comet outbursts. For new comets, especially, the explosive release of energy from free radicals in a rapidly rotating nucleus might produce splitting.

As pointed out in the discussion, several complicated chemical effects may occur in the nucleus of a comet as a result of radioactivity. Additional free radicals might result. On the other hand, Cameron notes that radioactive heating will tend to release chemical energy from radicals as they recombine into more stable molecules. This could be a very significant heat source, comparable to or greater than that of the possible radioactivity itself. The combined heat sources plus migration of volatiles might serve to strengthen the chemical bonds of the earthy material to produce sizable meteoroids in comets originating in the interstellar fragmented clouds.

From a contrary point of view, comets produced in the outer planetary cloud might, in later stages, acquire a coating of exotic molecules from the last fall-in of interstellar material. Hence, we should watch for layered comets, especially among new ones such as Kohoutek (1973f). Possibly, the parent molecules change after a new comet passes perihelion and loses a number of meters of its outer radius.

Comets appear to be as individual as human beings; indeed, neglecting size variations, I suspect that they vary much more in character. Thus, we need precise data on the dust-to-gas ratio for each comet. It is commonly stated that new comets tend to be dusty and old periodic comets to be gassy. I question the truth of this assumption because of the large meteoritic contribution by periodic comets. I suspect that the meteoritic material is simply more tightly bound together near the centers of large cometary nuclei and that the dust-to-gas ratio may actually be larger but unobserved for periodic comets. This could result from the radioactive heating or from original accretion.

Better stated, we need a sound taxonomic classification of comets on the basis of physical characteristics and spectroscopic information. We may be dealing with a mixture of sources that includes both those formed in the outer planetary regions and those formed at much greater solar distances.

Many of us are working with NASA toward the objective of unmanned space missions to comets and to asteroids (Stuhlinger, 1972; Whipple, 1973). There is some real hope of a flyby to Comet Encke in 1980. A study shows that the amount of scientific data expected in cometary missions varies inversely as the relative velocity at closest approach. Of course, the information does not go to infinity at zero velocity in a rendezvous mission, but does increase tremendously. Thus, an enormous gain will occur in the scientific value of such missions if they can utilize solar-electric propulsion. This allows the matching of spacecraft velocity with comets and asteroids and also permits more multiple missions to different objectives.

Some rather simple data, such as the density of the nucleus, could give strong diagnostic value as to the origin of comets. The conclusion by John S. Lewis that a frozen-methane nucleus would have a density ~ 1.0 g cm^{-3}, vs. 1.7 g cm^{-3} for a water-ice nucleus, is highly important. A still higher mean density would imply a considerable earthy center. Incidentally, a landing on a comet nucleus might be extremely difficult, because the outflowing gases would repel the spacecraft with a force greater than that of the gravitational attraction. Comet nuclei will probably have to be harpooned!

In a similar vein, a good measurement of the mean density of an asteroid would, in itself, have strong diagnostic value.

Now I would like to return to questions of particle size in comets. We observe two characteristic sizes, a maximum mass concentration in the zodiacal cloud near 10^{-5} g (Whipple, 1967) and another concentration in the micron-to-submicron region both in comets and in the zodiacal cloud. The latter seems to be in the size range where Field must place the lost refractories in the interstellar clouds, his conclusion being consistent with those in the papers in this volume by Salpeter, Grossman, and Wood.

No mention has been made of the magnetic properties of these fine grains. Aviva Brecher reports in discussion that the magnetic grains found in carbonaceous chondrites are also present in the submicron region. They appear to show chemical magnetic remanence and to be of primitive origin at very low temperature. They lose both their magnetic and their chemical character if heated only to 120 K. Thus, they must have been formed in fields of the order of 1 G and have never been heated appreciably. Cameron points out that their ages, 4.5×10^9 years, can refer only to the time of their accumulation, which is essentially the same as the chondrites as a whole, and not necessarily to the time of their actual crystallization. This leaves open the possibility that they are really more primitive than the solar system, i.e., of interstellar origin, or that they formed in the very outer reaches of the proto-solar cloud. Brecher notes that formation of the magnetite by oxidation should produce nickel-rich rims, which are not present. Cameron also notes that if the magnetite grains were produced at extremely low pressures, following Wood's discussion, their mode of origin is again inconsistent with the inner solar nebula, but they could have formed in higher magnetic fields. Grossman confirms this opinion, as he cannot reconcile the magnetite grains with Edward Ander's suggestion that they formed in the inner solar nebula. The partial pressure in iron at 400 K is negligible, with very little iron present.

Woolf also shows that the grains could not have formed in a circumstellar cloud, since the magnetic field could not exceed 10^{-3} G; he does not feel that pressure plays a significant role in their formation. Greenberg points out that one could not replace silicate cores for interstellar grains with magnetite and satisfy cosmic abundances. Thus, they could not constitute more than ~1% of the cores.

McCrosky finds that the dust from bright fireballs does not add information about cometary grains, because the atmospheric collections from other than the Revelstoke event are mostly contamination. In that case, the grain size is determined for the most part by filter characteristics.

Relating this discussion to cometary solids, I conclude that the 10^{-5}-g maximum does not yet appear to be connected with recognized grain growth processes. In other words, the maximum may result from some facet of accumulation, on the one hand, or some aspect of the destructive processes in interplanetary space, on the other. The micron or submicron particles of comets and the zodiacal light, however, may well be connected genetically with interstellar grains. Hemenway's solar grains present another possibility for the fine dust in the zodiacal cloud, unrelated either to comets or to the interstellar medium. No evidence exists for such extremely refractory fine particles in meteorites, at least to my knowledge, so it might appear that Hemenway's process was not a significant contributor to the solar nebula.

As a last comment, I am delighted with Herbig's recent evidence for great luminosity outbursts in young stars. In the late 1940s, I felt that the removal of a primitive atmosphere from the early earth must have required a truly violent proto-solar event. I called it "the bath of fire" and wondered whether proto-Mercury may not have been immersed in the proto-sun for a short time, possibly long enough to remove a considerable silicate mantle. Such an event could fit well with the new lunar data. Venus does not show any effect intermediate between those of earth and Mercury, so I do not want to press the idea. On the other hand, violent outbursts in the young sun could well have been effective in removing the embarrassingly large mass of gaseous solar nebula from the planetary regions.

Finally, thank you all again for being here and contributing so much both to the scientific content of this symposium and to the scientific comradeship.

REFERENCES

Bertaux, J. L., and Blamont, J., 1970. Compte Rendu Acad. Sci. Paris, Ser. B 270, 1518.

Biermann, L., and Trefftz, E., 1964. Zeits. für Astrophys. 59, 1.

Cameron, A. G. W., 1962. Icarus 1, 13.

Code, A. D., Houck, T. E., and Lillie, C. F., 1972. In The Scientific Results from the Orbiting Astronomical Observatory (OAO-2), ed. by A. D. Code, NASA SP-310, p. 109.

Cook, A. F., Jacchia, L. G., and McCrosky, R. E., 1963. Smithsonian Contr. Astrophys. 7, 209.

Delsemme, A. H., 1972. In Comets, Scientific Data and Missions, ed. by G. P. Kuiper and E. Roemer (Lunar and Planetary Laboratory, Univ. Arizona, Tucson), p. 32.

Delsemme, A. H., and Swings, P., 1952. Ann. d'Astrophys. 15, 1.

Donn, B., and Urey, H., 1956. Astrophys. Journ. 123, 339.

Everhart, E., 1973. Astron. Journ. 73, 329.

Gossner, J. L., and Whipple, F. L., 1949. Astrophys. Journ. 109, 380.

Hamid, S. E., 1950. Doctoral thesis, Harvard University.

Hamid, S. E., Marsden, B. G., and Whipple, F. L., 1968. Astron. Journ. 73, 727.

Hamid, S. E., and Whipple, F. L., 1953. Astron. Journ. 58, 100.

Harwit, M., 1967. In The Zodiacal Light and the Interplanetary Medium, ed. by J. L. Weinberg, NASA SP-150, p. 307.

Harwit, M., M., 1968. Astrophys. Journ. 151, 789.

Huebner, W., and Weigert, A., 1966. Zeits. für Astrophys. 64, 185.

Jacchia, L. G., 1955. Astrophys. Journ. 121, 521.

Kiang, T., 1972. Mem. Roy. Astron. Soc. 76, 27.

Marsden, B., 1968. Astron. Journ. 73, 367.

Marsden, B., 1969. Astron. Journ. 74, 720.

Marsden, B., 1970. Astron. Journ. 75, 75.

Marsden, B. G., and Sekanina, Z., 1971. Astron. Journ. 76, 1135.

Marsden, B. G., and Sekanina, Z., 1973. Astron. Journ. 78, 1118.

Marsden, B. G., and Sekanina, Z., 1974. Astron. Journ. 79, 413.

McCrea, W. H., 1960. Proc. Roy. Soc. London A256, 245.

McCrosky, R. E., 1955. Astron. Journ. 60, 170.

McCrosky, R. E., 1958. Astron. Journ. 63, 97.

McCrosky, R. E., Posen, A., Schwartz, G., and Shao, C.-Y., 1971. Journ.
    Geophys. Res. 76, 4090.

Oort, J. H., 1950. Bull. Astron. Inst. Netherlands 11, 91.

Öpik, E., 1932. Proc. Amer. Acad. Arts Sci. 67, 169.

Öpik, E. J., 1951. Proc. Roy. Irish Acad. 54, 165.

Öpik, E. J., 1965. Mem. Soc. Roy. Sci. Liège, Ser. 5 12, 523.

Öpik, E. J., 1966. Irish Astron. Journ. 7, 141.

Öpik, E. J., 1971. Irish Astron. Journ. 10, 35.

Porter, J. G., 1963. In The Moon, Meteorites and Comets, ed. by B. M. Middlehurst
    and G. P. Kuiper (Univ. Chicago Press, Chicago), Chap. 16.

Southworth, R. B., 1963. Smithsonian Contr. Astrophys. 7, 299.

Stuhlinger, E. (chairman), 1972. Report of the Comet and Asteroid Mission Study Panel,
    NASA TM-X-64677, 93 pp.

Whipple, F. J. W., 1930. Journ. Roy. Meteorol. Soc. 56, 287.

Whipple, F. L., 1948. Sci. Amer. 178, 34.

Whipple, F. L., 1950. Astrophys. Journ. 111, 375.

Whipple, F. L., 1955. Astrophys. Journ. 121, 750.

Whipple, F. L., 1962. Astronautics 7, 40.

Whipple, F. L., 1964. Proc. Nat. Acad. Sci. 52, 565.

Whipple, F. L., 1966. Science 153, 54.

Whipple, F. L., 1967. In The Zodiacal Light and the Interplanetary Medium, ed. by
    J. L. Weinberg, NASA SP-150, p. 409.

Whipple, F. L., 1972. In From Plasma to Planet, ed. by A. Elvius (John Wiley &
    Sons, New York), p. 211.

Whipple, F. L., chairman, 1973. The 1973 Report and Recommendation of the NASA
    Science Advisory Committee on Comets and Asteroids, NASA TM-X-71917, 75 pp.

Whipple, F. L., and Stefanik, R. P., 1966. Mem. Soc. Roy. Sci. Liege, Ser. 5 12,
    33.

Whipple, F. L., and Hamid, S. E., 1951. Helwan Obs. Bull. No. 41, 1.

Wright, F. W., and Whipple, F. L., 1950. Harvard Coll. Obs. Tech. Rep. No. 6, 44 pp.

APPENDIX

FRED LAWRENCE WHIPPLE

PERSONAL:

Born: Red Oak, Iowa, November 5, 1906; moved to Long Beach, California, January 1922; to Cambridge, Massachusetts, September 1931

Son of Harry Lawrence Whipple and Celestia (MacFarland) Whipple

Married Dorothy Woods of Los Angeles, California, 1928; divorced 1935

One Son — Earle Raymond

Married Babette Frances Samelson, Cambridge, Massachusetts, 1946

Two Daughters — Dorothy Sandra and Laura

EDUCATION:

1923	Long Beach, California, High School
1923—24	Occidental College
1924—27	A.B., University of California at Los Angeles
1927—31	Ph.D., University of California at Berkeley

HONORARY DEGREES:

1945	M.A., Harvard University
1958	D.Sc., American International College
1961	D.Sc., Temple University
1961	D.Litt., Northeastern University
1962	LL.D., C.W. Post College of Long Island University

POSITIONS HELD:

1927—29	Teaching Fellow, University of California
1929, summer	Instructor, Stanford University
1930—31	Lick Observatory Fellow
1931, summer	Instructor, University of California
1931—	Staff Member, Harvard College Observatory
1932—37	In charge of Oak Ridge Station, Harvard College Observatory
1932—38	Instructor, Harvard University
1938—45	Lecturer, Harvard University
1942—45	Research Associate, Radio Research Laboratory, O.S.R.D., in charge of development of confusion reflectors, "Window," as a radar countermeasure; was sent twice on missions to the United Kingdom and once to Mediterranean Theatre, 1944; "Window" was used extensively by the American Air Force.

POSITIONS HELD (cont.):

1945–50	Associate Professor, Harvard University
1947–49	Chairman, Committee on Concentration in the Physical Sciences, Harvard University
1949–56	Chairman, Department of Astronomy, Harvard University
1950–	Professor, Harvard University
1955–73	Director, Smithsonian Institution Astrophysical Observatory
1968–	Phillips Professor of Astronomy

COMMITTEES, OFFICES (United States):

U.S. House of Representatives
    Special Consultant, Committee on Science and Astronautics, 1960–1973

Gordon Research Conference on the Chemistry and Physics of Space
    Chairman, 1963

National Aeronautics and Space Administration
    Director, Optical Satellite Tracking Project, 1958–1973
    Project Director, Orbiting Astronomical Observatory, 1958–1972
    Director, Meteorite Photography and Recovery Program, 1962–1973
    Consultant, Aeronomy Subcommittee, 1961–1963
    Consultant, Planetary Atmospheres, 1962–1963
    Member, Space Sciences Working Group on Orbiting Astronomical Observatories, 1959–1969
    Member, Working Group on Geodetic Satellite Program, 1963–1968
    Member, Optical Astronomy Panel, Astronomy Missions Board, 1968–1970
    Member, Science and Technology Advisory Committee, 1969–1970
    Member, Comet and Asteroid Science Advisory Committee, 1971–1972; Chairman, 1972–1973
    Co-chairman, Comets and Asteroid Science Seminar, 1973–1974

International Geophysical Year
    Chairman, Technical Panel on Rocketry, 1955–1959
    Member, Technical Panel on Earth Satellite Program, 1955–1959
    Member, Working Group on Satellite Tracking and Computation, 1955–1958
    Chief Investigator, Project Optical Tracking of Artificial Earth Satellites, 1955–1958

National Academy of Sciences, National Research Council
    Member, Committee on Meteorology, 1958–1961
    Member, Bio-Astronautics Committee of the Armed Forces, 1959–1961
    Member, Space Science Board, 1958–1963
      Member, Committee on Potential Contamination and Interference from Space Experiments (Subcommittee of Space Science Board), 1963–
    Member, Committee on Atmospheric Sciences, 1958–1962
    Participant, Space Science Summer Study, 1962

U.S. Air Force
    Member, Scientific Advisory Board, 1953–1962
      Geophysics Panel, Space Technology Panel
    Associate Advisor, 1963–1967

COMMITTEES, OFFICES (United States) (cont.):

National Science Foundation
  Member, Advisory Panel on Astronomy, 1952—1955
  Chairman, 1954—1955
  Member, Divisional Committee for Mathematical and Physical Sciences, 1964—1968

Rocket and Satellite Research Panel
  Member, 1946—1958

National Advisory Committee on Aeronautics
  Member, Subcommittee, 1946—1952

Research and Development Board Panel
  Member, 1947—1952

University Corporation for Atmospheric Research, Boulder, Colorado
  Trustee-at-large, 1964—1967
  Member, Committee on NCAR Staff-University Relations, 1965—1968

Harvard Meteor Research Project
  Active leader of project on Upper Atmospheric and Meteor Research via meteor
    photography, sponsored by
  Bureau of Ordnance, U.S. Navy, 1946—1951
  Air Research and Development Command, U.S. Air Force, 1948—1962
  Office of Naval Research, 1951—1957
  Office of Ordnance Research, U.S. Army, 1953—1957

Harvard Radio Meteor Project
  Project Director, sponsored by
    National Bureau of Standards, Dept. of Commerce, 1957—1961
    National Science Foundation, 1960—1963
    National Aeronautics and Space Administration, 1963—1965

COMMITTEES, OFFICES (International):

International Astronomical Union
  Member, Commission 6, 1955—1970
    Vice President, 1961—1964
    President, 1966—1970
  Member, Commission 15, 1952—
    Vice President, 1961—1964
    President, 1964—1967
  Member, Commission 20, 1932—1958
  Member, Commission 22, 1946—
    President, 1946—1952
  President, Sub-Commission 22a, 1955—1961
  Member, Commission 34, 1950—1955
  Voting Representative, U.S.A., 1952, 1955
  Member, Commission 36, 1935—1952
  Member, Commission 44, 1959—1967

International Scientific Radio Union, National Committee (U.S.A.)
  Member, Commission 3, 1949—1961

COMMITTEES, OFFICES (International) (cont.):

Committee on Space Research (COSPAR), 1960–1968
    Member, Working Group on Geodetic Satellites
    Member, Working Group on Tracking, Telemetry, and Dynamics
    Chairman, Scientific Council on the Geodetic Uses of Artificial Satellites,
      1965–1968

International Astronautical Federation
    Member, 1955–

International Academy of Astronautics, 1961–
    Member, Scientific Advisory Committee, 1962–1966

Royal Society of Sciences, Liege, Belgium
    Corresponding Member, 1962–

Inter-American Astrophysical Congress, Mexico
    Delegate, 1942

RECIPIENT:

Donohue Medals for the independent discovery of six new comets

Presidential Certificate of Merit for scientific work during World War II

J. Lawrence Smith Medal of the National Academy of Sciences for research on meteors,
    1949

Exceptional Service Award by U.S. Air Force for scientific service in conjunction with
    Scientific Advisory Board, 1960

Medal from the University of Liege, Belgium, for astronomical research, 1960

Space Flight Award, American Astronautical Society, 1961

Commander of the Order of Merit for Research and Invention, Esnault-Pelterie award
    by the Ministry of Education, Public Health and Industry, France, 1962

Distinguished Federal Civilian Service Award from President Kennedy, June 1963

Space Pioneers Medallion for contributions to Federal Space Program, 1968

NASA Public Service Award for contributions to OAO 2 development, May 1969

NASA Public Service Group Achievement Award, for significant contribution to the
    success of the Apollo Program, October 1969

Leonard Medal of the Meteoritical Society, 1970

Career Service Award, from the National Civil Service League, 1972

The Henry Medal, presented by the Board of Regents of the Smithsonian Institution,
    1973

SOCIETY MEMBERSHIPS:

American Astronomical Society
    Vice-President, 1948–1950
    Member, Division for Planetary Sciences, 1969–

SOCIETY MEMBERSHIPS (cont.):

American Astronautical Society (Fellow)
  Vice-President, 1962–1964

American Geophysical Union (Fellow)
  Member, Upper Atmosphere Committee, Meteorology Section, 1957–1962
  Member, Committee on Cosmic and Terrestrial Relationships, 1957–1962

American Institute of Aeronautics and Astronautics (formerly Institute of Aerospace
    Sciences)
  Member, Aerospace Technology Panel for Space Physics, 1960–1963

American Rocket Society (Fellow)
  Member, Physics of the Atmosphere and Space Committee, 1959–1961

American Standards Association
  Member, Committee on Standardization in the Field of Photography, 1938–1950

National Academy of Science, 1959–

Royal Society of Arts, London
  Benjamin Franklin Fellow, 1968–

Associate, Royal Astronomical Society, 1970–

American Academy of Arts and Sciences, 1941–
  Councilor, Section on Astronomy and Earth Science, 1966–1968

American Philosophical Society
  Councilor, 1966–1969
  Committee on the Magellanic Premium, 1966
    Chairman, 1968–
  Chairman, Committee on Nomination of Officers, 1968–1969

Cosmos Club of Washington, D.C.

Examiner Club of Boston, Massachusetts

Phi Beta Kappa

Pi Mu Epsilon

Sigma Xi

Shop Club, Harvard University

American Association for the Advancement of Science

American Meteoritical Society
  Councilor, 1968–1970

American Meteorological Society

Astronomical Society of the Pacific

The Marquis Biographical Library Society (Advisory Member)

INVENTIONS:

Tanometer
Meteor Bumper

EDITORIAL POSITIONS:

Harvard Announcement Cards
    Editor, 1952—1960

Astrophysical Journal
    Associate Editor, 1952—1954

Astronomical Journal
    Associate Editor, 1954—1956; 1964—1971

Smithsonian Contributions to Astrophysics
    Editor, 1956—1973

Planetary and Space Science
    Regional Editor, 1958—

Space Science Reviews
    Editorial Board, 1961—

Annual Review of Astronomy and Astrophysics
    Editorial Committee, 1965—1969

Earth and Planetary Science Letters
    Editorial Board, 1966—1973

Comments on Astrophysics and Space Physics
    Contributor, 1968—1970

AUTHOR:

Earth, Moon and Planets, Blakiston Company, 1942, rev. 1963, rev. 1968, Harvard University Press; and Survey of the Universe with D. H. Menzel and G. de Vaucouleurs, Prentice-Hall, 1970. Scientific papers on orbits of comets, asteroids, and meteors; spectrophotometry of Cepheid variable stars, novae, and supernovae; colors of external galaxies; planetary nebulae; earth's upper atmosphere and nature of meteors by two-camera photographic method; interstellar medium; stellar and solar system evolution; theory of comets; optical tracking of artificial earth satellites; astronomy from space stations; meteoritical impact and erosion in space. Various subjects in the Encyclopaedia Britannica and various popular articles on astronomical subjects.

INDEX